Quantitative Applications in the Social Sciences

A SAGE PUBLICATIONS SERIES

Quantitative Applications in the Social Sciences

A SAGE PUBLICATIONS SERIES

Series/Number 07–152

MODERN METHODS
FOR ROBUST REGRESSION

Robert Andersen
University of Toronto

SAGE Publications
Los Angeles ▪ London ▪ New Delhi ▪ Singapore

For information:

Sage Publications, Inc.
2455 Teller Road
Thousand Oaks, California 91320
E-mail: order@sagepub.com

Sage Publications India Pvt. Ltd.
B 1/I 1 Mohan Cooperative Industrial Area
Mathura Road, New Delhi 110 044
India

Sage Publications Ltd.
1 Oliver's Yard
55 City Road
London EC1Y 1SP
United Kingdom

Sage Publications Asia-Pacific Pte. Ltd.
33 Pekin Street #02-01
Far East Square
Singapore 048763

Library of Congress Cataloging-in-Publication Data

Andersen, Robert, Ph.D.
Modern methods for robust regression / Robert Andersen.
    p. cm.—(Quantitative applications in the social sciences; 152)
Includes bibliographical references and index.
ISBN 978-1-4129-4072-6 (pbk.)
    1. Social sciences—Statistical methods. 2. Regression analysis. 3. Robust statistics. I. Title.
HA31.3.A63 2008
519.5′36—dc22

                                    2007009019

07   08   09   10   11   10   9   8   7   6   5   4   3   2   1

Acquisitions Editor:	Vicki Knight
Associate Editor:	Sean Connelly
Production Editor:	Melanie Birdsall
Copy Editor:	Liann Lech
Typesetter:	C&M Digitals (P) Ltd.
Proofreader:	Kevin Gleason
Indexer:	Sheila Bodell
Cover Designer:	Candice Harman
Marketing Manager:	Stephanie Adams

CONTENTS

LIST OF FIGURES

LIST OF TABLES

SERIES EDITOR'S INTRODUCTION

In 1886, Francis Galton published the groundbreaking article titled "Regression Towards Mediocrity in Hereditary Stature" that began the development of the statistical method of linear regression as we know it today. Upon analyzing data for 205 parents and 928 children, Galton found that taller or shorter parents tended to have children who were not as tall or short, a property summarized by the statistical term of "regression toward the mean."

To show how regression works with such human height data, I use a set of similar data, attributed to Galton's disciple Karl Pearson, that is sex-specific. The figure below plots the height (in inches) of 1,078 pairs of fathers and sons, and the data, indicated by the circles, clearly follow a linear trend portraying the phenomenon of regression toward the mean (i.e., 45).

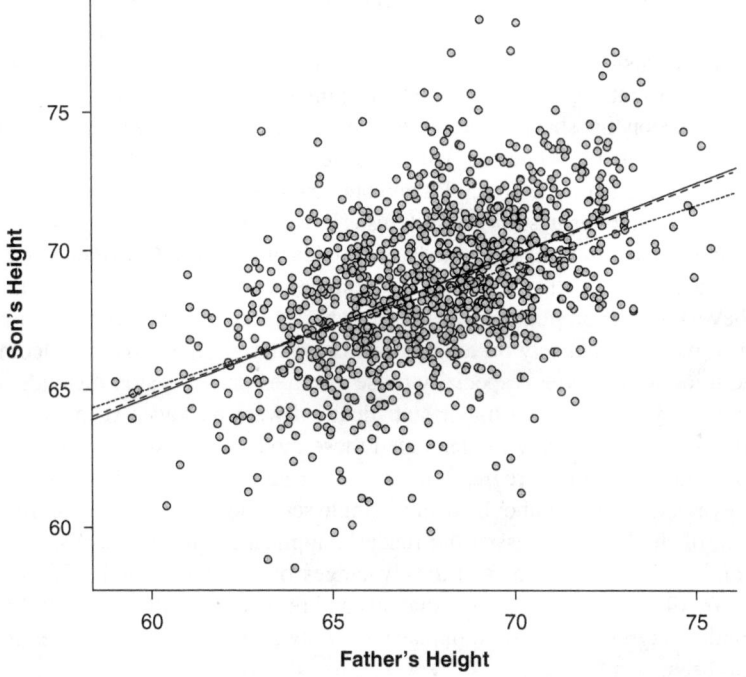

In the figure, I fit a regression line, indicated by the solid line, with a slope estimate of 0.514, obtained by ordinary least squares estimation (all two-tailed tests for this and later estimates are much more significant than the conventional 0.001 level, thus not reported here). By anyone's standard, the data are quite well behaved. However, even in such well-behaved data, some cases are more deviant than others: One can quickly discern that there are a few cases located in the upper-right and the lower-left quadrants of the figure that seem to be away from the majority of the cases that center around the regression line. Were these cases more to the extreme, one could apply one of the standard quick "fixes" that include exclusion from the analysis, recoding (if coding error), and addition of new variables to the analysis. But if there are no justifiable fixes available for dealing with these unusual (or not so unusual) cases, what can a data analyst do? This is where robust and resistant regression methods come in.

To demonstrate robust and resistant regression, I fit two more lines to the data (by using R's MASS package), the dashed line indicating the regression line (slope estimate = 0.502) estimated by robust regression using an MM-estimator and the dotted line indicating the regression line (slope estimate = 0.442) estimated by resistant regression via the minimization of quantile squared residuals (where largest ones are ignored). One can easily see that an estimation of robust regression using the MM-estimator produced a slope that is only slightly less steep than OLS regression. However, resistant regression obtained an estimate that differs more, and gives a conclusion that shows a greater degree of regression toward the mean. The current book by Andersen will help social scientists understand these methods and learn the principles and applications of robust regression, with the focus on robustness of validity (instead of efficiency).

In the social sciences, the modern methods of robust and resistant regression are not yet well known. (These methods are termed "modern" because they are often computation intensive, a character of many of today's statistical methods that rely on today's fast computers.) This book, a welcome addition to the series, especially those volumes on regression methods, is extremely timely when major statistical software packages such as SAS and Stata have already implemented these modern regression methods. It presents many robust regression methods from different sources and how they relate to one another by using a single set of notations, a nice contribution of the book. To assist the reader's application, it also discusses the comparative advantages and disadvantages of various methods. Through books like this, social science students and researchers will eventually find modern regression as commonplace and easy to use as classical regression has been.

—Tim F. Liao
Series Editor

ACKNOWLEDGMENTS

I thank John Fox; reviewers Kenneth A. Bollen, University of North Carolina, and Jeff Gill, University of California, Davis; and the QASS series editor Tim F. Liao for their helpful critiques of earlier drafts of the book. I also thank participants in my ICPSR Regression III course for their suggestions on how to improve the lecture notes that formed the start of the book. Finally, thanks to Jill, Kamma, Lene, Hanne, and Kristina for being patient while I finished the book.

MODERN METHODS FOR ROBUST REGRESSION

Robert Andersen
University of Toronto

1. INTRODUCTION

Regression analysis is the workhorse of statistical methods in quantitative social science. A vast majority of problems are tackled by either linear models or generalized linear models. When properly fitted, regression estimates provide a powerful and elegant summary of relationships in the data. On the other hand, if conducted blindly and mechanically, regression analysis can lead to erroneous conclusions. One cause for concern is the presence of unusual observations, which can sometimes severely distort estimates from ordinary least squares (OLS) regression, even when the data set is large. Although less common, unusual observations can also cause havoc for generalized linear models. This underscores the importance of detecting and properly handling outliers in regression analysis.

There is much merit in incorporating "modern" regression methods, such as nonparametric regression, as diagnostic tools within the general linear model and generalized linear model frameworks (see, for example, Cook and Weisberg 1999; Fox 1997; Hastie, Tibshirani, and Friedman 2001). Referred to as "modern" because they rely on intense computation to fit many regressions to calculate final estimates, these methods can uncover a whole range of problems—especially nonlinearity, but also other problems when applied to residuals—that are often difficult to uncover using only OLS estimation. Only recently, as personal computers have become much faster, have social scientists become aware of the benefits of these methods.

Robust regression is another set of modern techniques that tend to be computationally intensive. If there is no fix for unusual cases—that is, if they can't be recoded or dealt with by a transformation or the addition of new terms to the model—robust regression is a suitable alternative to OLS. The many types of robust regression have in common the goal of providing unbiased estimates that are unaffected by outliers or skewed residual distributions. The most useful of these methods also try to produce relatively efficient estimates when the errors are normally distributed. At the very least, robust regression techniques are useful diagnostic tools for uncovering potentially problematic cases.

Although robust regression is not widely used in the social sciences, statisticians have known its benefits for many decades, and they continue to develop new methods. Recently, a number of good monographs geared toward statisticians discuss these developments (see, for example, Atkinson and Riani 2000; Lawrence and Arthur 1990; Myers 1990; Maronna, Martin, and Yohai 2006; Wilcox 2005). On the other hand, aside from a few good general articles on robust regression (e.g., Western 1995; Dietz, Frey, and Kaloff 1987; Wu 1985), the differences between a large number of types of robust regression have yet to be explored in a forum accessible to practicing social scientists. The present book attempts to rectify this situation.

The goal of this book is to discuss various methods for detecting and properly handling influential cases in regression analysis. Chapter 2 defines some terms important to understanding the robustness of an estimator. Because they form the basis of robust regression techniques, it also deals with various measures of location and scale. Chapter 3 outlines the ways that unusual observations and skewed distributions can affect OLS estimates. It also gives an overview of some traditional techniques—both formal statistical tests and graphical methods—for detecting influential cases in the general linear model. Chapter 4 discusses various robust regression methods for linear models and their limitations. Chapter 5 provides a discussion of standard errors for robust regression estimates, focusing largely on bootstrapping. Chapter 6 briefly describes the generalized linear model and some diagnostics for detecting unusual observations within it. More important, it extends robust regression to the generalized linear model. Chapter 7 summarizes the book and provides some general advice on how to deal with unusual observations. Finally, the appendix summarizes facilities for robust regression in some commonly used statistical computer programs.

The book is organized with the research process in mind: Once we have fit a model to test our hypotheses, what can we do to ensure that there are no problematic cases that make our inferences and tests inaccurate? Also, how do we handle such cases if they exist? Before tackling these larger issues, it is important to define the general concept of robustness—and more specifically, robust and resistant regression—and to motivate the book with a real-world example.

Defining Robustness

As Huber (2004:1) states, "The word 'robust' is loaded with many—sometimes inconsistent—connotations." Nonetheless, it is generally accepted that when evaluating an estimator, two types of robustness should be considered.

These are summarized by Mosteller and Tukey (1977:203–209), who argue that a robust estimator satisfies two conditions: (1) if a small change is made to the data, it will not cause a substantial change in the estimate, and (2) the estimate is highly efficient under a wide range of circumstances. The first condition, which reflects the *resistance* of the estimator to unusual observations, can be considered *robustness of validity*. In other words, the estimator provides a valid estimate for the bulk of the data. The second condition, which relates to the underlying distributional assumptions, can be referred to as *robustness of efficiency*. This condition implies that a failure to meet the distributional assumptions of the estimator has little impact on its precision (i.e., on its standard errors).

This book is primarily concerned with *robustness of validity*—that is, the level of resistance to change that an estimator has to unusual observations. It will also address *robustness of efficiency,* although this is regarded as a secondary criterion for a robust estimator. That is, after the description of the robustness of validity for an estimator, there will be some discussion of its efficiency. It should also be made clear that although conceptually distinct, skewed distributions and outliers have similar consequences for an estimator, in terms of both robustness of validity and robustness of efficiency. This book's main focus, however, is on the impact of outliers. In short, it compares the robustness of a wide variety of estimators that attempt to limit the influence of unusual observations. In this respect, the terms *resistance* and *robustness* will be treated as synonymous throughout the book—they refer to how much impact individual observations have on the estimator.

Defining Robust Regression

Not surprisingly, "robust" also takes on many meanings with respect to regression analysis. One definition has to do with so-called *robust standard errors,* which are used to account for some pattern of heteroscedasticity or error dependency. Although useful, these do not fit the definition of robust regression upon which this book rests. As said above, the present book is concerned with methods that explicitly try to accommodate—or perhaps better said, discount—unusual observations. In this regard, two definitions of robust regression are directly relevant.

The first considers all regression models that explicitly accommodate heavy-tailed error distributions and outliers as robust regression models. The second distinguishes between *robust regression* and *resistant regression.* According to this definition, robust regression techniques are concerned with both robustness of validity and robustness of efficiency. These techniques use information from all observations but give less weight to those that are highly unusual. Various types of robust regression consider unusual residuals, discrepancy in terms of the independent variables (leverage), or a combination of the two. Most of these methods give reasonably efficient estimates, both when the error distribution has heavy tails and when it is normal.

Methods typically referred to as resistant regression, on the other hand, often have little concern for efficiency. The main goal of these methods is to prevent unusual observations from drastically influencing the values of the regression slopes. Rather than only down-weight unusual observations, resistant methods often set criteria for completely removing them from the analysis.

It is also common to distinguish robust regression methods with a high breakdown point from those with a low breakdown point. Yet another classification divides models according to whether or not they have bounded influence. These concepts will be defined in detail in Chapter 2; for now it is sufficient to say that a very robust estimator has bounded influence and a high breakdown point. At one time these distinctions were important, but most new developments in robust regression combine a high breakdown point with bounded influence.

For purposes of the present book, then, robust regression is defined broadly as any regression that limits the influence of unusual observations on the values of its estimates. As a result, the various regression techniques will not be separated into "robust" and "resistant" categories. Instead, the book explores various methods in terms of how they developed in response to previously existing methods, although all of the criteria above will be discussed with respect to each method.

A Real-World Example: Coital Frequency of Married Couples in the 1970s

It is perhaps well known that unusual observations can cause problems for regression estimates if the sample size is small, and this makes sense intuitively as well. If n is small, there are few cases to counteract a largely discrepant case. On the other hand, it is more difficult for discrepant observations to radically alter the regression surface in large data sets because there are many observations to counteract the discrepant case. Using simple regression as an example, we know that the regression line is found by minimizing the sum of the squared residuals $\sum e_i^2$, where the e_i are simply the differences between the observed values of y_i and the values predicted by the regression $\hat{y}_i$ ($e_i = y_i - \hat{y}_i$). If there is only one discrepant observation among several thousands, it is difficult for that observation to pull the regression line toward it because so many counteracting residuals must be kept small. This does not mean that discrepant cases cannot wreak havoc on regression estimates for large data sets, however. The following is a good example of how they can.

Using panel data from the National Fertility Studies, research by Jasso (1985) estimated age and period effects, controlling for cohort effects, on the monthly coital frequency of married couples from 1970–1975. Her major findings were (1) controlling for cohort and age effects, there was a negative period effect, and (2) controlling for period and cohort effects, wife's age had a positive effect. These findings differed significantly from previous findings and, not surprisingly, resulted in publication in the *American Sociological Review*.

To Jasso's credit, her discussion of the methods she employed was sufficiently clear for other researchers to be able to replicate her results. Kahn and Udry (1986) did just that, raising several concerns with Jasso's original analysis. Firstly, they claimed that four cases were seemingly miscoded a value of 88. They argued that these values were likely supposed to be coded as 99, the value assigned to missing information in the data set. They based this conclusion on the fact that no other respondent reported a value greater than 63, and 99.5% of the observations had values less than 40. Second, using model diagnostics, Kahn and Udry detected an additional four outliers. Each of these respondents had a much lower value in the previous wave of the study, suggesting that the outliers were not typical and thus could be justifiably removed. Finally, they argued that Jasso missed an interaction between length of marriage and wife's age. Kahn and Udry's reanalysis of the data—which included dropping the eight outliers (from a sample of more than 2,000) and adding the interaction—drastically changed the findings. As Table 1.1 indicates, when the outliers are removed, Jasso's new findings appear a lot less important. More specifically, the effect of wife's age (which was logged to account for a nonlinear pattern) is no longer statistically significant at standard levels.

In her rejoinder, Jasso (1986) argued that Kahn and Udry's analysis generates a new problem of "sample truncation bias." She claimed that by removing the outliers, Kahn and Udry had artificially confined the dependent variable to a segment of its range. She objects to this practice, arguing that researchers should not remove unusual observations simply because they fail to conform to their beliefs. She further claims that it is reasonable to expect that a large random national sample would select a few widely discrepant observations because there are large differences in coital frequencies according to culture and region.

TABLE 1.1
Determinants of Marital Coital Frequency

	Model 1	Model 2	Model 3	Model 4
Period	−.72***	−.67***	−3.06**	−0.08
Log Wife's Age	27.61**	13.56	29.49	−1.62
Log Husband's Age	−6.43	7.87	57.89	−5.23
Log Marital Duration	−1.50***	−1.56***	−1.51*	1.29
Wife Pregnant	−3.71	−3.74***	−2.88***	−3.95*
Child Under 6	−0.56**	−0.68***	−2.91***	−0.55**
Wife Employed	0.37	0.23	0.86	0.02
Husband Employed	−1.28**	−1.10**	−4.11***	−0.38
R^2	0.0475	0.0612	0.2172	0.0411
n	2062	2055	243	1812

SOURCE: Adapted from Kahn and Udry (1986: table 1).

NOTES: Model 1: Jasso's original analysis; Model 2: four "miscodes" and four other outliers dropped; Model 3: Marital duration $\leq$ 2 years (excluding miscodes and outliers); Model 4: Marital duration > 2 years (excluding miscodes and outliers).
*$p < .05$; **$p < .01$; ***$p < .001$.

This example illustrates three important points. First, it shows the value of using diagnostic tools to uncover potentially problematic observations. Second, it shows how outliers can influence regression estimates even in large data sets. Third, the debate reflects the fact that there is no universally accepted method for handling unusual observations. The decision on what action should be taken when influential observations are detected should be based on substantive knowledge. In other words, the researcher must make a judgment call. With respect to this particular example, I leave it to those with better substantive knowledge of the topic to decide the best way to handle the outliers. It is sufficient for the purpose of this book to show that, despite a sample of more than 2,000 observations, as few as eight outliers drastically altered the results. If one decides that the observations should not be ignored, they can be handled by removing them, as did Kahn and Udry, or by robust regression.

2. IMPORTANT BACKGROUND

We now turn to various concepts important to assessing the robustness of an estimator. In this regard, bias, consistency, efficiency, breakdown point, and the influence function will be defined. All of these will be used throughout the book.

Bias and Consistency

Assume a sample, Z, with n observations. Let $T_n(Z_1, \ldots, Z_n)$ with probability distribution P represent an estimator for the parameter θ. In other words, applying T to Z gives the estimate of the population parameter:

$$T(Z) = \hat{\theta} \qquad [2.1]$$

The estimator is unbiased if

$$E[T(Z)] = E(\hat{\theta}) = \theta. \qquad [2.2]$$

In other words, the *average of an unbiased statistic equals the population parameter.* It follows, then, that the bias of an estimator $T(Z) = \hat{\theta}$ is given by

$$\text{bias } E[T(Z) - \theta]. \qquad [2.3]$$

Unbiasedness is certainly important, but consistency is also of concern when determining the "best" estimator to use. *An estimator $\hat{\theta}$ is consistent if it converges to θ as the sample size grows to infinity.* We can also consider consistency in terms of the mean squared error (MSE) of an estimate. In this respect, $\hat{\theta}$ is consistent if

$$\lim_{n \to \infty} \text{MSE}(\hat{\theta}) = 0. \qquad [2.4]$$

Breakdown Point

The breakdown point (BDP)[1] is a *global measure of the resistance* of an estimator. More specifically, it is the smallest fraction or percentage of discrepant data (i.e., outliers or data grouped at the extreme end of the tail of the distribution) that the estimator can tolerate without producing an arbitrary result (Hampel 1974; Huber 2004). Assume all possible "corrupted" samples Z' that replace m observations in the data set with arbitrary values (i.e., observations that do not fit the general trend in the data). The maximum effect[2] that could arise from these substitutions is

$$\text{effect } (m; T, Z) = \frac{\sup}{Z'} \|T(Z') - T(Z)\|, \qquad [2.5]$$

where the supremum is over all possible Z'. If the effect $(m; T, Z)$ is infinite, the m outliers have an arbitrarily large impact on T. In other words, the estimator "breaks down" and fails to adequately represent the pattern in the bulk of the data. More generally, the breakdown point for an estimator T for a finite sample Z is defined as

$$BDP(T, Z) = \min\left\{\frac{m}{n} : \text{effect } (m; T, Z) \text{ is infinite}\right\}. \qquad [2.6]$$

The highest possible breakdown point for an estimator is 50%, which indicates that as many as half the observations could be discounted. A breakdown point higher than 0.5 is undesirable because it would mean that the estimate could pertain to less than half of the data. The goal of a robust estimator is to sufficiently capture the pattern in the bulk of the data. In other words, a breakdown point greater than zero is a desirable attribute. In fact, Hampel et al. (1986) argue that data sets typically contain as much as 10% of observations that deviate from the general pattern characterized by the bulk of the data, suggesting that a robust estimator should have a breakdown point of at least 10%. As we shall see later, however, some of the first proposed robust regression estimators have a breakdown point of 0 or very close to it.

Influence Function

Originally proposed by Hampel (1974; see also Hoaglin, Mosteller, and Tukey 1983:350–358; Jurečková and Picek 2006:27–32), the *influence function* of an estimator measures the impact of a single observation y_i that contaminates the theoretically assumed distribution F of an estimator T. In other words, whereas the breakdown point measures global robustness, the influence function (IF) measures *local resistance or, more specifically, infinitesimal perturbations* on the estimator. Also referred to as the influence curve (or sensitivity curve when viewed with respect to a single sample), the influence function for an estimator T is defined by

$$IF(Y, F, T) = \lim_{\lambda \to 0} \frac{T[(1 - \lambda)F + \lambda\delta_y] - T(F)}{\lambda}, \qquad [2.7]$$

where δ_Y is the point of contamination at y (i.e., at y and 0 otherwise) with probability mass λ. In other words, λ gives the proportion of contamination at y. Simply put, the IF indicates the change in an estimate caused by adding arbitrary outliers at the point y, standardized by the proportion of contamination.

A bounded influence function is a desirable attribute of a robust estimator because it means that the influence of a particular observation can get only so high. An unbounded influence function allows the influence of "contaminated" observations to continue to grow, regardless of how unusual they are. In other words, there is no limit on the effect of discrepancy. As we shall see later, the influence function for OLS regression is unbounded and proportional to the size of the residual, meaning that a highly discrepant residual can completely destroy the OLS estimator. Many early robust regression methods also have unbounded influence functions, resulting in a resistance that is sometimes no better than that of OLS. Most robust estimators commonly employed today, however, have both a high breakdown point and bounded influence function.

Relative Efficiency

Another important concept for understanding robust estimation is efficiency. If the goal is to make inferences about a larger population from sample data, we desire an unbiased estimator that is as efficient as possible. In the strictest sense, the efficiency of an estimator is determined by the ratio of its minimum possible variance to its actual variance. Only when the ratio is equal to one—that is, when it has the lowest possible variance—is an estimate considered efficient.[3] An estimator is asymptotically efficient if it reaches efficiency with large samples. More generally, an estimator is considered to be efficient if its sampling variance is relatively small, resulting in small standard errors. It follows that some estimators are more efficient than others, and thus the concept of *relative efficiency* is useful for assessing competing estimators.

For most kinds of estimation, there is one estimator that has maximum efficiency under some particular assumptions. We can use this estimator as a benchmark to which we compare the efficiency of other estimators. Assume that we have two estimators T_1, and T_2, for the population parameter θ. If T_1 has maximum efficiency and T_2 is less efficient, T_1 will also have a smaller mean squared error. The relative efficiency of T_2 is determined by the ratio of its mean squared error to the mean squared error for T_1:

$$\text{Efficiency } (T_1, T_2) = \frac{E\left[(T_2 - \theta)^2\right]}{E\left[(T_1 - \theta)^2\right]} \qquad [2.8]$$

If the assumptions of linearity, constant error variance, and uncorrelated errors are met, OLS estimates are the most efficient of unbiased linear estimators. As a result, relative efficiency of robust estimators is assessed in comparison to the OLS estimators under these conditions. Although no robust regression estimator is more efficient than OLS under these conditions, several estimators are nearly as efficient, and at the same time have the desirable property of high resistance to outliers. The relative efficiency of robust regression estimators should be considered cautiously, however, because it is asymptotic efficiency that is typically assessed (Ryan 1997:354). In other words, relative efficiency is meaningful only with sufficiently large sample sizes. Little is known about the small sample properties of most robust regression estimators, resulting in the common practice of using bootstrapping to find standard errors in these situations.

Measures of Location

Although there are various types of regression, all predict conditional values of a dependent variable from some predictor(s) by taking into account some measure of location and scale of the response variable. OLS, for example, estimates the conditional mean of a dependent variable y from one or more independent variable xs. Because OLS is based on the mean, which is not resistant to outliers, its estimates can also be affected by outliers. Similarly, estimates from generalized linear models (GLMs) are not completely resistant to outliers because they estimate the conditional mean of a linear predictor. Robust regression methods rely on more robust measures of location and/or scale. It is helpful, then, to discuss various measures of location and scale before exploring the regression techniques that use them.

A measure of *location* is a quantity that characterizes a position in a distribution. Typically, measures of center are of most concern, although other measures of location (quantiles, for example) can also be considered. Assume a random variable Y with distribution F. An estimate $\theta(Y)$ is a measure of location of F if, for any constants a and b, four conditions[4] are met (Wilcox 2005:20–21):

a. $\theta(Y + a) = \theta(Y) + a$
b. $\theta(-Y) = -\theta(Y)$
c. $Y \geq \theta$ implies $\theta(Y) \geq 0$
d. $\theta(bY) = b\theta(Y)$

Condition (a), which requires that when a constant is added to all values of Y, the measure of location will increase by the same amount, is referred to as *location equivariance*. Taken together, Conditions (a), (b), and (c) require that the value of the measure is within the range of Y. Condition (d) means that the measure has *scale equivariance*. In other words, if all values of Y are multiplied by a particular value (i.e., if the scale is altered), the measure of location will be altered by the same factor.

The Mean

The most common measure of location is the mean. Consider independent observations y_i and a simple model estimating the center μ of a population distribution

$$y_i = \mu + e_i, \qquad [2.9]$$

where the e_i represent the residuals. If the underlying distribution is normal, the sample mean is the maximally efficient estimator of μ, producing the fitted model

$$y_i = \bar{y} + e_i. \qquad [2.10]$$

Despite its widespread use, including in OLS regression, the mean is not a robust measure of location. If the distribution has heavy tails or outliers, the mean is less efficient than many other measures of center and, more important, can often be misleading. Even the addition of a single badly miscoded observation can alter its estimate.

Consider the following five observations for the variable y:

$$y_1 = 3 \quad y_2 = 3 \quad y_3 = 4$$
$$y_4 = 5 \quad y_5 = 5$$

Applying the well-known formula for the sample mean, $\bar{y} = \frac{1}{n} \sum_{i=1}^{n} y_i$, produces $\bar{y} = 4$. We now replace just one of the observations, y_3, with a "bad" observation (assume that it is a miscode), giving the following values of y:

$$y_1 = 3 \quad y_2 = 3 \quad y_3 = 44$$
$$y_4 = 5 \quad y_5 = 5$$

For these new data, $\bar{y} = (3 + 3 + 44 + 5 + 5)/5 = 12$. The mean has been dramatically pulled toward the outlier, taking a value three times larger than when the outlier is excluded. In fact, the "contaminated" mean is much larger than any of the observed values *except* the "bad" observation.

Because even a single observation can cause the mean to break down, its breakdown point is $BDP = \frac{1}{n}$, and thus effectively 0 when n is large. Just as problematic, the influence of each observation on the mean is proportional to the size of y. The mean is found by minimizing the least squares objective function:

$$\sum_{i=1}^{n} (y_i - \hat{\mu})^2 = 0 \qquad [2.11]$$

Taking the derivative with respect to y produces the influence function

$$IF_{\bar{y}}(y) = 2y. \qquad [2.12]$$

This, of course, is not an attractive attribute for data that are not "well behaved" (i.e., that have outliers or a heavy tail).

One strategy to combat the influence of outliers on the mean is to use a two-step procedure, where the outliers are first identified and removed before calculating the mean. Rather than calculating the mean for the distribution excluding the outliers, Hampel (1974) shows that using a robust measure of location is usually a better way to proceed. Many measures of location are less vulnerable than the mean to outliers. In other words, many estimators are more robust.

α-Trimmed Mean

A relatively robust measure of center is the *trimmed mean,* which reduces the impact of outliers or heavy tails by removing the observations at the tails of the distribution. Let $y_1, \ldots y_n$ represent observations on a variable from a random sample. We start by ordering the values of y from lowest to highest, $y_{(1)} \leq y_{(2)} \leq \cdots \leq y_{(n)}$, and determining the desired amount of trimming, $0 = \alpha < 0.5$. the mean is then calculated for all observations *except* the g smallest and largest observations $g = [\alpha n]$, where $[\alpha n]$ is rounded to the nearest integer. The formula for the trimmed mean can be written as[5]

$$y_t = \frac{y_{(g+1)} + \cdots + y_{(n-g)}}{n - 2g}. \qquad [2.13]$$

The breakdown point of the trimmed mean is determined by the amount of trimming, and thus is $BDP = \alpha$. A simple rule of thumb is to remove 10% of the observations from each tail of the distribution (i.e., set $\alpha = 0.2$). Leger and Romano (1990) further suggest calculating the mean for $\alpha = 0, 0.1$, and 0.2 and choosing the value that gives the lowest standard error for the final calculation. The amount of trimming also determines the influence function. Unlike for the mean, the influence for the trimmed mean is bounded, although there are marked increases at y_α and $y_{1-\alpha}$.[6] Its influence function can be written as

$$IF_{\bar{y}_t}(y) = \begin{cases} \frac{y_\alpha - \hat{\mu}_t}{1 - 2\alpha} & \text{for } y < y_\alpha \\[2mm] \frac{y - \hat{\mu}_t}{1 - 2\alpha} & \text{for } y_\alpha \leq y \leq y_{1-\alpha} \\[2mm] \frac{y_{1-\alpha} - \hat{\mu}_t}{1 - 2\alpha} & \text{for } y > y_{1-\alpha} \end{cases} \qquad [2.14]$$

where $\hat{\mu}_t$ is the trimmed mean (see Wilcox 2005:29). The relative efficiency of the trimmed mean depends on the distribution. If the distribution is normal and too much trimming is done, precision will be reduced because it results in greater spread relative to the smaller n, thus increasing the estimate of the spread of its sampling distribution. On the other hand, if the distribution has heavy tails and extreme outliers, trimming can result in improved efficiency because the variance of y—and hence the estimated variance of the sampling distribution of its mean—is decreased. Judgments on the amount of trimming should be made only after careful examination of the distribution.

The Median

The *median M* is simply the value of y that occupies the middle position when the data are ordered from smallest to largest. To find the median, we start by ordering the observations from lowest to highest value, $y_{(1)} \leq y_{(2)} \leq \cdots \leq y_{(n)}$. The median is given by

$$M = y_{((n+2)/2)} \quad \text{if } n \text{ is an odd number}$$

and

$$M = .5y_{(n/2)} + .5y_{(n/2+1)} \quad \text{if } n \text{ is an even number.}$$

Equivalently, the median minimizes the absolute values objective function

$$\sum_{i=1}^{n} |y_i - \hat{\mu}| = 0. \qquad [2.15]$$

Taking the derivative of Equation 2.15 gives the shape of the influence function

$$IF_M(y) = \begin{cases} 1 & \text{for } y > 0 \\ 0 & \text{for } y = 0 \\ -1 & \text{for } y < 0. \end{cases} \qquad [2.16]$$

As the bounded influence function indicates, the median is highly resistant to outliers. Its robustness is also reflected in its breakdown point of $BDP = 0.5$. The disadvantage of the median is that it has relatively low efficiency compared to the mean when the distribution is normal. In these situations, the sampling variance for the mean is σ^2/n, whereas the sampling variance for the median is $\pi/2 = 1.57$ times larger at $\pi\sigma^2/2n$ (Kenney and Keeping 1962:211).

Measures of Scale

Let Y represent a random variable. A measure of scale is any nonnegative functional $\tau(Y)$ that satisfies the following conditions (Wilcox 2005:34)[7]:

a. The measure is *scale equivariant*, meaning that $\tau(aY) = a\tau(Y)$, where a is a constant that is greater than 0.
b. The measure is *location invariant*, meaning that $\tau(Y + b) = \tau(Y)$, where b is a constant.
c. The measure is *sign invariant*, $\tau(Y) = \tau(-Y)$.

There are too many measures of scale to include them all, so we concentrate on those that are most relevant to robust regression. We explore mostly how outliers affect the magnitude of the scale estimate, paying little attention to efficiency. For more discussion on the latter, see Wilcox (2005).

Standard Deviation

The most commonly employed measure of scale is the standard deviation s, which is defined by

$$s_y = \sqrt{\frac{\sum_{i=1}^{n} (y_i - \bar{y})^2}{n-1}}. \qquad [2.17]$$

If the distribution of y is normal, this is the most appropriate measure of scale because of its superior efficiency. On the other hand, the standard deviation is not robust to heavy-tailed distributions or distributions with outliers. Because it is based on the mean—which has an unbounded influence function and breakdown point of 0—the standard deviation inherits these qualities. As a result, robust regression techniques typically use other measures of scale.

Mean Deviation From the Mean

The mean deviation from the mean (MD), sometimes known more simply as the *mean deviation*, is given by

$$MD = \frac{\sum_{i=1}^{n} |y_i - \bar{y}|}{n}. \qquad [2.18]$$

The MD is relatively efficient compared to the standard deviation when the distribution of y has heavy tails, but it also has the undesirable property of a breakdown point of 0 and an unbounded influence function. Although important for some early robust regression techniques, the MD should generally be seen as obsolete given that there are now much more robust measures of scale.

Mean Deviation From the Median

The mean deviation from the median, MDM, is a slight improvement over the MD in terms of robustness. Rather than find the absolute difference of y from the mean, the MDM finds the absolute differences from the median M, resulting in

$$MDM = \frac{\sum\limits_{i=1}^{n} |y_i - M|}{n}.$$ [2.19]

Although it also uses the median, MDM still relies on mean deviations, and thus has a breakdown point of $BDP = 0$ and an unbounded influence function (see Wilcox 2005:35 for more details). In other words, the mean deviation from the median is not immune to extreme outliers and heavy tails, and thus it is not ideal for use in robust regression.

Interquartile Range

The q-quantile range QR_q is a set of bounded influence measures of scale that can have a very high breakdown point. Any particular q-quantile range is given by

$$QR_q = y_{1-q} - y_q, \text{ where } 0 < q < .5.$$

Setting $q = .25$ (i.e., the difference between the .25 and .75 quantiles) produces the interquartile range (IQR), which, with a breakdown point of $BDP = 0.25$, is the most robust and thus most commonly used of the quantile ranges (Wilcox 2005:35–36). The influence function for the IQR is given by the influence function at the third quartile minus the influence function at the first quartile (i.e., $IF_{.75} - IF_{.25}$):

$$IF_{IQR}(y) = \begin{cases} \frac{1}{f(y_{.25})} - C & \text{if } y < y_{.25} \text{ or } y > y_{.75} \\ -C & \text{if } y_{.25} \leq y \leq y_{.75} \end{cases}$$ [2.20]

where

$$C = q \left\{ \frac{1}{f(y_{.25})} + \frac{1}{f(y_{.75})} \right\}$$ [2.21]

The high breakdown point and bounded influence function of the IQR are desirable properties, leading to its use in some early robust regression techniques. It still plays a role in quantile regression, which will be introduced later. There are more robust measures of scale, however, so despite its simplicity, the IQR is seldom incorporated in more recent developments in robust regression.

Median Absolute Deviation

The median absolute deviation (MAD) is defined by

$$MAD = \text{median} |y_i - M|.$$

Based entirely on variation around the median, the MAD is far more resistant to outliers than the standard deviation and measures of absolute deviation associated with the mean.[8] The MAD achieves the highest possible breakdown point of $BDP = 0.5$ and has a bounded influence defined by

$$IF_{MAD}(y) = \frac{\text{sign}(|y - M| - MAD) - \frac{f(M+MAD) - f(M-MAD)}{f(M)} \, \text{sign}(y - M)}{2[f(M+MAD) + f(M-MAD)]} \qquad [2.22]$$

where $f(y)$ is the probability density function for y (see Wilcox, 2005:35 for more details). An attractive attribute of the MAD is that it can be adjusted to ensure consistency for large sample sizes under the assumptions that $y \sim N(\mu, \sigma^2)$ by multiplying by 1.4826 (approximately $1/\Phi^{-1}(3/4)$, where Φ is the normal probability density function). All of these attributes make the MAD an attractive measure of scale for robust regression, at least as an initial estimate.

M-Estimation

M-estimation includes a large class of estimators that generalize the idea of maximum likelihood to robust measures of scale and location (Huber 2004). M-estimation is also the foundation for many robust regression estimates, including those classified as M-estimates, GM-estimates, S-estimates, and MM-estimates. All of these will be discussed in Chapter 4. When formulated properly, M-estimates are very robust, especially with respect to estimating location. They are also relatively efficient compared to other robust measures for large samples ($n \geq 40$), becoming more efficient as n gets larger (Hogg 1974; see also Wu 1985).

Assume that $y_1, \ldots, y_n$ is independently and identically distributed according to $F(y; \theta)$. Let $T_n(y_1, \ldots, y_n)$ be an estimate of an unknown parameter θ that characterizes the distribution $F(y; \theta)$. The likelihood of the estimator is given by

$$L(\theta; y_i, \ldots, y_n) = \prod_{i=1}^{n} f(y; \theta), \qquad [2.23]$$

where $f(y;\theta)$ is the probability density function corresponding to $F(y;\theta)$. The maximum likelihood estimator is the value of θ that maximizes the likelihood function or, equivalently, minimizes the objective function $\rho(y;\theta)$:

$$-\log l = \sum_{i=1}^{n} \rho(y;\theta) \qquad [2.24]$$

Restricting the objective function $\rho(y;\theta)$ to any function that is differentiable with an absolutely continuous derivative $\Psi(.)$ results in the maximum likelihood estimator T_n,

$$\sum_{i=1}^{n} \Psi(y;\theta) = 0, \qquad [2.25]$$

where

$$\Psi(y;\theta) = -(\partial/\partial\theta)\rho(y;\theta)$$
$$= (\partial/\partial\theta)\log f(y;\theta). \qquad [2.26]$$

In order for the maximum likelihood estimator—or M-estimator—to be uniquely determined, $\rho(y;\theta)$ must be strictly convex, and thus the score function $\Psi(y;\theta)$ must be strictly increasing. Using $\rho(y;\theta) = -\log f(y;\theta)$ gives the ordinary maximum likelihood estimate (see Huber 2004: chap. 3).

M-estimates take on many different forms, the properties of which are determined by the choice of $\rho(.)$ or, equivalently, $\Psi(.)$. If $\Psi(.)$ is unbounded, the breakdown point of the estimator is $BDP = \lim_{n\to\infty} BDP = 0$. Conversely, if $\Psi(.)$ is odd and bounded, and thus $\rho(.)$ is symmetric around 0, the breakdown point of the estimator is $BDP = 0.5$. The score function $\Psi(.)$ has the same shape as the influence function proposed by Hampel (1974). More specifically, $IF(y;F,T) = \Psi(y)/\gamma(F)$, where $\gamma(F) = \int f(y)\,d\Psi(y)$. The proportionality constant $[\gamma(F)]^{-1}$ depends on both Ψ and the probability density function $f(y)$. In other words, the IF is the negative of the score function (see Jurečková and Sen 1996; Hoaglin et al. 1983:356).

M-Estimation of Location

Consider the population mean μ as the expectation of the random variable Y. Let $\rho(y - \hat{\mu})$ be an objective function that measures distance from an estimate of location $\hat{\mu}$. The M-estimate is found by minimizing the objective function

$$\sum_{i=1}^{n} \rho(y;\theta) = \sum_{i=1}^{n} \rho\left(\frac{y_i - \hat{\mu}}{cS}\right), \qquad [2.27]$$

where S is a measure of scale of the distribution and c is a tuning constant that adjusts the degree of resistance of the estimator by defining the center and tails of the distribution. Although M-estimates are location equivariant, *they are not scale equivariant* and thus the tuning constant is required. The smaller the value of c, the greater the resistance the estimate has to outliers.

Taking the derivative of Equation 2.27 gives the shape of the influence function. The M-estimator is then the value of $\hat{\mu}$ that solves

$$\sum_{i=1}^{n} \Psi\left(\frac{y - \hat{\mu}}{cS}\right) = 0. \qquad [2.28]$$

The measure of scale and the measure of location are estimated simultaneously, and thus an iterative estimation procedure is required (see Huber 2004 for extensive details). More details of estimation will be given with respect to M-estimates for regression in Chapter 5. For now, we continue with a general explanation extending from the mean.

M-estimation of the mean relies on the least squares objective function

$$\rho(y; \theta) = \frac{1}{2}(y - \hat{\mu})^2. \qquad [2.29]$$

The derivative of Equation 2.28 shows that influence is proportional to the value of y

$$\Psi(y; \theta) = (y - \hat{\mu}). \qquad [2.30]$$

To compute a more robust M-estimate than the mean, we simply replace the least squares objective function with another function that gives less weight to extreme values. The Huber weight function and biweight functions are two common choices.

Huber Estimates

At the center of the distribution, the Huber weight function behaves like the mean and the least squares objective function associated with it (i.e., observations are given equal weight), but at the extremes it behaves like the median, and the least absolute values objective function associated with it, giving decreasing weight to observations as they get farther out on the tails:

$$\rho_{\mathrm{H}}(y; \theta) = \begin{cases} \frac{1}{2}y^2 & \text{if } y \leq c \\ c|y| - \frac{1}{2}c^2 & \text{if } y > c \end{cases} \qquad [2.31]$$

Because the goal is to produce an estimate that is resistant to outliers, the MAD is typically used to calculate the measure of scale, S. Defining $S = MAD/0.6745$ results in S estimating σ when the population is normally distributed.

Following Huber (1964), it is convention (and standard in statistical software) to set $c = 1.345$, which gives substantial resistance to outliers ($1.345/0.6745 \cong 2\text{MADs}$) and produces a relative efficiency of approximately 95%.

Taking the derivative of Equation 2.31 gives the shape of the influence function

$$\Psi_H(y; \theta) = \begin{cases} c & \text{if } y > c \\ y & \text{if } y \leq |c| \\ -c & \text{if } y < -c. \end{cases} \qquad [2.32]$$

Finally, the derivative of $\Psi(.)$ gives the weights that are given to individual observations:

$$w_{H_i}(y) = \begin{cases} 1 & \text{if } y \leq c \\ c/|y| & \text{if } y > c \end{cases} \qquad [2.33]$$

Biweight Estimates

The major difference between the bisquare weight, also referred to as Tukey's bisquare, and the Huber weight occurs at the extreme ends of the tails of the distribution, where the biweight objective function is somewhat more resistant to outliers

$$\rho_{BW}(y; \theta) = \begin{cases} \frac{c^2}{6} \left\{ 1 - \left[1 - \left(\frac{y}{c}\right)^2\right]^3 \right\} & \text{if } |y| \leq c \\ \frac{c^2}{6} & \text{if } |y| > c. \end{cases} \qquad [2.34]$$

A tuning constant of $c = 4.685$ results in $4.685 \times S \cong 7\text{MADs}$, which produces 95% efficiency when sampling from a normal population (Huber 1964). Taking the derivative of Equation 2.34, we see that the influence function tends rapidly toward zero

$$\Psi_{BW}(y; \theta) = \begin{cases} y\left[1 - \left(\frac{y}{c}\right)^2\right]^2 & \text{if } |y| \leq c \\ 0 & \text{if } |y| > c. \end{cases} \qquad [2.35]$$

Taking the derivative of Equation 2.35 gives the weight function

$$w_{BW_i}(y) = \begin{cases} \left[1 - \left(\frac{y}{c}\right)^2\right]^2 & \text{if } |y| \leq c \\ 0 & \text{if } |y| > c. \end{cases} \qquad [2.36]$$

Figure 2.1 displays the Huber and biweight functions with their default tuning constants, applied to a uniform distribution ranging from -10 to

10. We see that the two M-estimators behave much more similarly to each other than they do to the mean, which gives all observations equal weight. The Huber and biweight functions work in a similar manner for most of the distribution, except in the very center and at the extreme tails. For the biweight function, all observations with an absolute value greater than five, $|y_i| > 5$, are given a weight of zero, and only observations directly in the middle receive a weight of one. On the other hand, the Huber weight gives none of the observations a weight of zero, and a significantly larger proportion of observations a weight of one.

Although the Huber weight function and the biweight function are the most commonly used in M-estimation, there are many other options, some of which are shown in Table 2.1. For more details about these estimators, especially regarding recommendations for the tuning constants, see Andrews et al. (1972) and Ramsay (1977).

M-Estimators of Scale

It is relatively straightforward to extend M-estimation to estimation of scale (Wilcox 2005:92–98). Again, the general idea is to find a function that gives less weight to extreme observations. The general class of M-estimators of scale are defined by the asymptotic variance of the M-estimate of location

$$\zeta^2 = \frac{K^2 \tau^2 E\left[\Psi^2(Z_i)\right]}{\{E[\Psi'(Z_i)]\}^2}$$

$$Z_i = \frac{y_i - \mu_m}{cS}, \qquad\qquad [2.37]$$

where μ_m is the M-estimate of location, c is a positive tuning constant, S is the initial measure of scale typically set to the MAD, and Ψ is the score function. As with M-estimation of location, the Huber weight function and the biweight function are typical choices. Because it is used more often and has been shown to be more efficient, we concentrate on the latter, which results in the *biweight midvariance* (see Lax 1985).

The *biweight midvariance* is both efficient and highly resistant to outliers, achieving a breakdown point of approximately 0.5 (Hoaglin et al. 1983). It is defined by

$$\hat{\zeta}^2_{\text{bimid}} = \frac{\displaystyle\sum_{i:y_i^2 \leq 1} \left(y_i - M_y\right)^2 \left(1 - Z_i^2\right)^4}{\left[\displaystyle\sum_{i:y_i^2 \leq 1} \left(1 - Z_i^2\right)\left(1 - 5Z_i^2\right)\right]^2}, \qquad\qquad [2.38]$$

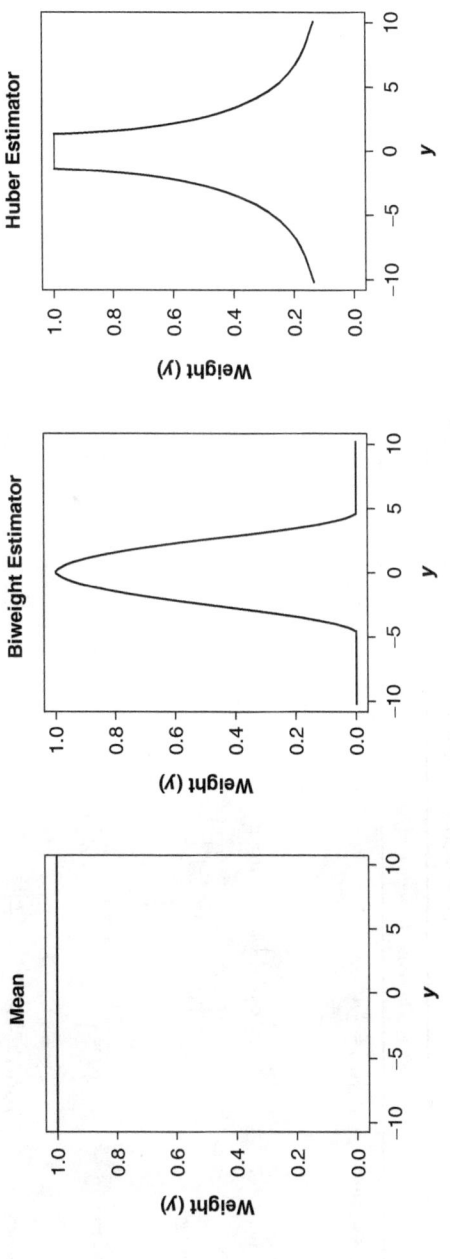

Figure 2.1 Commonly Used *M*-Estimator Weight Functions Compared to the Mean

TABLE 2.1
A Selection of Some Possible Functions for M-Estimation

	Objective Function, $\rho(\mu)$	Influence Function, $\psi(\mu)$	Weight Function, $w(\mu)$
Least Squares	$\rho_{LS}(\mu) = \frac{1}{2}\mu^2$	$\Phi_{LS}(\mu) = \mu$	$w_{LS}(\mu) = 1$
Huber	$\rho_H(\mu) = \begin{cases} \frac{1}{2}\mu^2 & \text{if } \mu \le c \\ c\lvert\mu\rvert - \frac{1}{2}m^2 & \text{if } \mu > c \end{cases}$	$\Phi_H(\mu) = \begin{cases} c & \text{if } \mu > c \\ \mu & \text{if } \mu \le c \\ -c & \text{if } \mu < -c \end{cases}$	$w_H(\mu) = \begin{cases} 1 & \text{if } \mu \le c \\ c/\lvert\mu\rvert & \text{if } \mu > c \end{cases}$
Biweight	$\rho_{BW}(\mu) = \begin{cases} \frac{c^2}{6}\left\{1 - \left[1 - \left(\frac{\mu}{c}\right)^2\right]^3\right\} & \text{if } \lvert\mu\rvert \le c \\ \frac{c^2}{6} & \text{if } \lvert\mu\rvert > c \end{cases}$	$\Phi_{BW}(\mu) = \begin{cases} \mu\left[1 - \left(\frac{\mu}{c}\right)^2\right]^2 & \text{if } \lvert\mu\rvert \le c \\ 0 & \text{if } \lvert\mu\rvert > c \end{cases}$	$w_{BW}(\mu) = \begin{cases} \left[1 - \left(\frac{\mu}{c}\right)^2\right]^2 & \text{if } \lvert\mu\rvert \le c \\ 0 & \text{if } \lvert\mu\rvert > c \end{cases}$
Andrew	$\rho_A(\mu) = \begin{cases} c\{1 - \cos(\mu/c)\} & \text{if } \lvert\mu\rvert \le c\pi \\ 2c & \text{if } \lvert\mu\rvert > c\pi \end{cases}$	$\Phi_A(\mu) = \begin{cases} \sin(\mu/c) & \text{if } \lvert\mu\rvert \le c \\ 0 & \text{if } \lvert\mu\rvert > c \end{cases}$	$w_A(\mu) = \begin{cases} \frac{\sin(\mu/c)}{(\mu/c)} & \text{if } \lvert\mu\rvert \le c \\ 0 & \text{if } \lvert\mu\rvert > c \end{cases}$
Ramsay	$\rho_R(\mu) = \frac{1 - e^{-c\lvert\mu\rvert}(1 + c\lvert\mu\rvert)}{c^2}$	$\Phi_R(\mu) = \mu e^{-c\lvert\mu\rvert}$ (maximum at c^{-1})	$w_R(\mu) = e^{-c\lvert\mu\rvert}$

SOURCE: Adapted from Draper and Smith (1998: table 25.1).

where M_y is the median of y and

$$Z_i = \frac{y_i - \mu_m}{cS}. \qquad [2.39]$$

It is important to note that the summation in the equation is restricted to $y_i^2 \leq 1$. The tuning constant, c, is typically set to 9 and the scale to MAD, resulting in maximum efficiency.

Comparing Various Estimates

EXAMPLE 2.1: Simulated Data

Table 2.2 compares the resistance of some of the estimators discussed thus far, applying them to simulated data. The first column applies the estimators to 20 random observations that were generated from the standard normal distribution $y_i \sim N(0, 1)$, having a range from -2.2 to 1.7. In other words, these data are well-behaved, containing no outliers. The second column applies the estimators to the same data but with the addition of an extreme outlier taking a value of 60, assumed to be a miscoded observation. The breakdown point of the estimators is shown in the third column.

The first panel of the table shows the results for various measures of location. Consistent with its $BDP = 0$, the mean is badly distorted by the outlier as it is pulled toward it (changing from 0 to 2.85). On the other hand, the trimmed mean—which, following convention, has trimmed 20% of the data from the tails and thus removed the outlier—has performed very well, taking on almost identical values for the good data and the contaminated data (-0.09 versus -0.04). The median and M-estimate (using bisquare weights), which both have BDP = 0.5, are also virtually unaffected by the outlier.

Turning now to the measures of scale, we see that those involving a mean in their calculation—that is, the standard deviation, the mean deviation from the mean, and the mean deviation from the median—are all badly distorted by the outlier. Of course, this is not surprising given that they all have $BDP = 0$. The standard deviation is most affected, taking on a value more than 13 times as large as it does in the absence of the outlier. On the other hand, the outlier has very little influence on the interquartile range ($BDP = 0.25$) and the median absolute deviation ($BDP = 0.5$), the two measures based on the median. Similar to the M-estimate of location, the outlier does not hinder the performance of the biweight midvariance, which has $BDP = 0.5$.

TABLE 2.2
Measures of Location and for Simulated Data
With and Without an Extreme Outlier

Estimator	Breakdown Point	All Observations, $\hat{\theta}_1$	Outlier Removed, $\hat{\theta}_2$
Measures of Location			
Mean	0	0	2.85
α-trimmed mean	α (proportion of trimming)	−0.09	−0.04
Median	.5	−0.02	0.005
M-estimation	.5	−0.12	−0.03
Measures of Scale			
Standard deviation	0	1	13.13
Mean deviation from mean	0	0.71	5.44
Mean deviation from median	0	0.61	2.89
Interquartile range	.25	1.07	1.21
Median absolute deviation	.5	0.61	0.66
Biweight midvariance	.5	0.89	1.06

EXAMPLE 2.2: Public Opinion Toward Pay Inequality in Cross-National Perspective

We now turn to an example using real social science data. The data in Table 2.3 are from Weakliem, Andersen, and Heath's (2005) cross-national study of the relationship between income inequality and public opinion on pay inequality. The data set contains information measured during the 1990s on 48 countries. The variables are as follows:

- *Secpay.* The average score on an item from the World Values Survey (Inglehart et al. 2000) that asked respondents their opinions about pay inequality (secpay). The wording of the question is as follows: "Imagine two secretaries, of the same age, doing practically the same job. One finds out that the other earns considerably more than she does. The better paid secretary, however, is quicker, more efficient and more reliable at her job. In your opinion, is it fair or not fair that one secretary is paid more than the other?" Respondents were given two response choices: "Fair" (coded 0), or "Not Fair" (coded 1). As a result, *high average scores reflect public opinion that favors equality* (i.e., a majority of respondents in the country answered that it was not fair for the two secretaries to have different salaries). The averaged score across countries ranges from 0.054 to 0.622 and has a mean of 0.2.

TABLE 2.3
Public Opinion and Economic and Political Variables for 48 Countries

Country	Public Opinion (Secpay)	Gini Coefficient	Per Capita GDP	Democracy
Armenia	.061	44.4	2072	0
Australia	.179	31.7	22451	1
Austria	.112	23.1	23166	1
Azerbaijan	.070	36.0	2175	0
Bangladesh	.057	28.3	1361	0
Belarus	.075	28.8	6319	0
Belgium	.302	27.2	23223	1
Brazil	.232	60.1	6625	0
Britain	.211	34.6	20336	1
Bulgaria	.164	30.8	4809	0
Canada	.176	28.6	23582	1
Chile	.361	56.5	8787	0
China	.131	41.5	3105	0
Croatia	.092	29.0	6749	0
Czech Republic	.557	26.6	12362	0
Denmark	.248	21.7	24217	1
Dominican Republic	.089	50.5	4598	1
Estonia	.054	35.4	7682	0
Finland	.354	22.6	20847	1
France	.231	32.7	21175	1
Georgia	.086	37.1	3353	0
Hungary	.115	28.9	10232	0
India	.226	29.7	2077	1
Ireland	.289	35.9	21482	1
Italy	.226	34.6	20585	1
Japan	.284	24.9	23257	1
Latvia	.070	28.5	5728	0
Lithuania	.096	33.6	6436	0
Mexico	.211	53.7	7704	0
Moldova	.127	34.4	1947	0
Netherlands	.328	31.5	22176	1
Norway	.441	24.2	26342	1
Peru	.175	46.2	4282	1
Portugal	.265	35.6	14701	1
Romania	.133	28.2	5648	0
Russia	.076	48.0	6460	0
Slovakia	.622	19.5	9699	0
Slovenia	.108	29.2	14293	0
Spain	.286	32.5	16212	1
Sweden	.401	25.0	20659	1
Switzerland	.149	36.1	25512	1
Taiwan	.075	27.7	12090	0

(Continued)

TABLE 2.3 (Continued)

Country	Public Opinion (Secpay)	Gini Coefficient	Per Capita GDP	Democracy
Turkey	.207	41.5	6422	0
Ukraine	.085	47.3	3194	0
Uruguay	.273	42.3	8623	0
USA	.148	36.9	29605	1
Venezuela	.208	46.8	5808	1
West Germany[a]	.149	30.0	22169	1

a. The survey was administered to respondents in West Germany only, and the data set uses the term "West Germany."

- *Gini.* The Gini coefficient, which theoretically ranges from 0 (perfect income equality where income is divided equally among all citizens) and 1 (perfect inequality, where one individual has all of the income). In other words, high values indicate high levels of income inequality.
- *Per Capita GDP/1000.* The per capita gross domestic product of the country in U.S. dollars.
- *Democracy.* A dummy variable coded 1 for "Old Democracies" (i.e., the country had experienced democratic rule for at least 10 years at the time of the data collection), and 0 for "New Democracies."

For more detailed information on the sources used to construct the measures, see Weakliem et al. (2005).

Of interest is the distribution of public opinion toward pay inequality (often referred to simply as "public opinion" from here onward) for those countries that were democratic for less than 10 years at the time of the study ($n = 26$). Given that the public opinion variable will be used as a dependent variable in regression analyses later, it is important to explore its distribution in a preliminary attempt to identify any features—such as a skew or outliers—that might be problematic. We start by examining Figure 2.2, which displays a kernel density estimate (i.e., a smoothed histogram) of the distribution of the public opinion variable. With the exception of a small bump at the extreme positive end of the distribution, the rest of the distribution is fairly symmetric. Further exploratory analysis indicates that two countries—the Czech Republic and Slovakia—have unusually high values. As can be seen in Table 2.4, these countries have values of 0.557 and 0.622, whereas no other country has a value reaching 0.4. Given that the two countries were joined until very recently, it seems likely that the uniqueness of these countries is due to a common cultural and historical heritage.

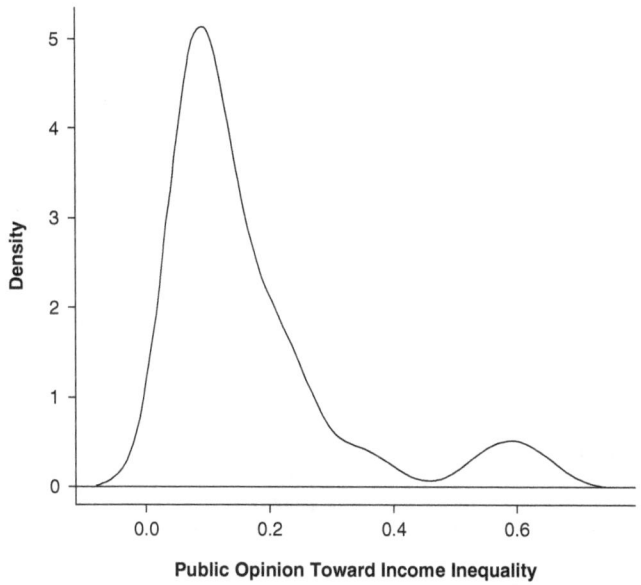

Figure 2.2   Distribution of Public Opinion Toward Pay Inequality for 26 New Democracies

We now turn to Table 2.4, which explores how various estimators of location and scale behave when the Czech Republic and Slovakia are included and excluded. Starting with the mean, we see that it decreases substantially (from 0.167 to 0.131) when the outliers are excluded. Similarly, the measures of scale based on the mean decrease substantially when the outliers are removed (e.g., the standard deviation is 1.86 times as large when the outliers are included than when they are excluded). On the other hand, the high resistance of the median and the M-estimate is evident in that they are virtually unchanged when the outliers are removed. Similarly, the differences in estimates between the two data sets are much smaller for the median absolute deviation and the M-estimate (biweight midvariance), two measures of location with high breakdown points.

In concluding this chapter, a cautionary note about examining the univariate distributions of the variables used in a regression analysis is appropriate. OLS regression estimates the *conditional mean* of y given the xs. As a result, an outlier for y is not necessarily a regression outlier. Conversely, it is not necessary that an influential observation in terms of the regression estimates is an outlier in terms of y. Still, this does not mean that we should

TABLE 2.4
Measures of Location and Scale for Public
Opinion Variable, New Democracies

Estimator	All Observations, $\hat{\theta}_1$	Czech Republic and Slovakia Removed, $\hat{\theta}_2$
Measures of Location		
Mean	0.167	0.131
α-trimmed mean	0.123	0.114
Median	0.112	0.102
M-estimation	0.127	0.112
Measures of Scale		
Standard deviation	0.145	0.078
Mean deviation from mean	0.102	0.060
Mean deviation from median	0.081	0.056
Interquartile range	0.129	0.097
Median absolute deviation	0.042	0.032
Biweight midvariance	0.005	0.004

ignore the univariate distributions. Failing to explore the univariate distributions could prevent the researcher from detecting important features of the data. But it is best to refrain from any remedies for the unusual observations until the relationships between the variables have been explored. With this in mind, we now turn to the OLS estimation of linear regression, exploring in detail how unusual observations can affect its estimates and how they can be detected. We return to measures of scale and location later in the context of robust regression methods.

Notes

1. Typically, ε_n^* is used instead of BDP to denote the breakdown point. I intentionally avoid the use of ε_n^* in order to prevent confusion with the errors for a regression model, which is unrelated to the breakdown point.

2. Although discussions of the breakdown point often use the term *bias,* the term *effect* is used here to avoid confusion with the usual statistical meaning of bias discussed earlier. If influential outliers do not reflect miscoding, an estimator can still be unbiased—that is, the average of the estimator from repeated random sampling will equal the population parameter—regardless of the effect the outliers have on the estimate. Still, this does not mean that the estimate will be a useful summary of the data.

3. If the estimator satisfies this condition, it is considered to have reached the Cramer-Rao lower bound (see Cramer 1946 for more details).

4. Bickel and Lehmann (1975) further suggest a fifth condition. Assume two random variables X and Y. If X is stochastically larger than Y—that is, for any value of x, $F_X(x) \leq F_Y(x)$—then to qualify as a measure of location, $\theta(X) \geq \theta(Y)$.

5. Although this is the most common definition of the trimmed mean, slightly different estimators have also been proposed (see Reed 1998; Kim 1992).

6. Huber (2004) shows further that the influence function for the trimmed mean can be derived even when the trimming is not symmetrical.

7. A measure of dispersion is a special measure of scale for which $\tau(X) \geq \tau(Y)$ when both X and Y have symmetric distributions but the stochastic distribution of $|Y|$ is larger than the stochastic distribution of $|X|$ (Bickel and Lehman 1976).

8. MAD is sometimes incorrectly used to refer to the much less robust "median deviation from the *mean*." The median deviation from the mean is of little use because when the conditions under which it should be used are met—a single-peaked, symmetric distribution—the standard deviation is more useful.

3. ROBUSTNESS, RESISTANCE, AND ORDINARY LEAST SQUARES REGRESSION

Ordinary Least Squares Regression

Indexing individual observations by i and variables by j, the linear regression model can be written as

$$y_i = \sum_{j=1}^{k} x_{ij}\beta_j + \varepsilon_i, \qquad [3.1]$$

where y_i is the dependent or response variable, the x_{ij} are independent variables predicting y, the β_j are the regression coefficients, and the ε_i represent a random component (i.e., errors) that is independent of the xs. The OLS solution minimizes the sum of squared residuals $\sum e_i^2$, which are estimates of the errors

$$\min \sum_{i=1}^{n} \left(y_i - \sum_{j=1}^{k} x_{ij}\beta_j \right)^2 = \min \sum_{i=1}^{n} e_i^2. \qquad [3.2]$$

Differentiating Equation 3.2 results in

$$\sum_{i=1}^{n} \left(y_i - \sum x_{ik}\beta_k \right) x_{ij} = \sum_{i=1}^{n} (e_i) x_{ij}. \qquad [3.3]$$

The residuals, which have expectation of 0, are defined simply as

$$e_i = y_i - \hat{y}_i, \qquad\qquad [3.4]$$

where the $\hat{y}_i$ are the fitted values of the model.

In matrix form, the linear regression model is

$$\mathbf{y} = \mathbf{X}\boldsymbol{\beta} + \boldsymbol{\varepsilon}, \qquad\qquad [3.5]$$

where $\underset{(n\times1)}{\mathbf{y}}$ is the vector of observations of the response variable, $\underset{(n\times k+1)}{\mathbf{x}}$ is the model matrix containing all values of the explanatory variables for each observation, $\underset{(k+1\times n)}{\boldsymbol{\beta}}$ is the vector of unknown coefficients, and $\underset{(n\times1)}{\boldsymbol{\varepsilon}}$ is the vector of random errors, which are simply differences between the observed values $\mathbf{y}$ and their expected values $E(\mathbf{Y})$.

The least squares solution for the coefficients is

$$\hat{\boldsymbol{\beta}} = \left(\mathbf{X}^{\mathsf{T}}\mathbf{X}\right)^{-1}\mathbf{X}^{\mathsf{T}}\mathbf{y}. \qquad\qquad [3.6]$$

The fitted or predicted values of the response variable are then derived from

$$\hat{\mathbf{y}} = \mathbf{X}\left(\mathbf{X}^{\mathsf{T}}\mathbf{X}\right)^{-1}\mathbf{X}^{\mathsf{T}}\mathbf{y} = \mathbf{H}\mathbf{y}, \qquad\qquad [3.7]$$

where $\mathbf{H}$ (the "hat matrix") is a symmetric $n \times n$ matrix that projects $\mathbf{y}$ onto the predicted values $\hat{\mathbf{y}}$. The diagonal elements of $\mathbf{H}$, known as *hat values* h_i, give the leverage of the observations. An observation has high *leverage* on the regression surface if it has an unusual x value—that is, it takes a value far from the mean of x. As we shall see later, the residuals and the hat values give important information about the amount of influence individual observations have on the regression coefficients.

Classical statistical inference for the linear model rests on some assumptions about the errors:

1. The dependent variable y has a linear relationship with the predictors. In other words, the expected value of ε given a particular value of x_i is 0, $E(\varepsilon_i) = 0$.
2. The errors have constant variance at all values of the xs, $V(\varepsilon_i|x_i) = \sigma_\varepsilon^2$. This is typically referred to as the homoscedasticity assumption.
3. The errors ε_i are uncorrelated, $\text{cov}(\varepsilon_i, \varepsilon_j) = 0$, for $i \neq j$.

According to the Gauss-Markov theorem, if these assumptions are met, the OLS estimators are the *best linear unbiased estimators* (BLUE) of the population regression coefficients (Draper and Smith 1998:136). In fact, $\hat{\boldsymbol{\beta}}$ provides unbiased estimates of minimum variance regardless of the distributional properties of the errors. If we add the assumption that the errors are normally distributed, and thus $\varepsilon_i \sim N\left(0, \sigma_\varepsilon^2\right)$, $\hat{\boldsymbol{\beta}}$ provides the maximum likelihood estimates of $\boldsymbol{\beta}$ and has an easily derivable probability distribution. More specifically, if the errors are normally distributed, the coefficients are also normally distributed because they are a linear function of the errors. Following from the central limit theorem, however, normality is crucial for the standard errors of OLS estimators only when n is small.

The variance-covariance matrix of the regression coefficients is given by

$$V\left(\hat{\boldsymbol{\beta}}\right) = \left(\mathbf{X}^\mathbf{T}\mathbf{X}\right)^{-1}\sigma_\varepsilon^2, \qquad [3.8]$$

where the variance of the errors σ_ε^2 is estimated by the variance of the residuals s_e^2. The standard errors of the coefficients are given by the square root of the diagonal terms in Equation 3.8. Large residuals—which reflect outliers or heavy tails—increase the estimate of s_e^2 and thus inflate the standard errors of the estimates. Outliers can also be associated with non-constant error variance, and thus the OLS estimates lose efficiency because they give equal weight to all observations including the outliers, despite that the latter contain less information about the regression. As we shall see later, unusual observations in the model matrix $\mathbf{X}$ also have important implications, good or bad depending on the type of unusualness, for the standard errors of OLS estimators (see Cook and Weisberg 1999:161).

Implications of Unusual Cases for OLS Estimates and Standard Errors

In order to understand how unusual observations can affect regression estimates, it is helpful to define four concepts: univariate outlier, regression outlier, leverage, and influence (Cook and Weisberg 1982; see also Rousseeuw and van Zomeran 1990; Fox 1991, 1997). A *univariate outlier* is an observation that stands away from the rest of the distribution of a particular variable. Although it is prudent to detect such cases in preliminary analyses before performing a regression analysis, they are not necessarily problematic. In other words, an observation that is unconditionally unusual in either its y value or x value is not necessarily an outlier in the regression.

A *regression outlier*, sometimes called a *vertical outlier* (Rousseeuw and van Zomeren 1990), stands apart from the general pattern for the bulk of the data. More specifically, it is an observation that is discrepant in terms of its y value, conditional on its value of x. Regression outliers are typically characterized by large residuals. A large residual *does not* necessarily mean that the observation affects the estimate of the regression slopes, however. Nor does a small residual, at least when from an OLS regression, necessarily indicate that an observation follows the pattern of the bulk of the data. When an outlier is influential, it pulls the regression surface toward it, and thus it can actually have a very small residual.

An observation has *leverage* on the regression surface if it has an unusual *x* value. More specifically, the further the observation sits from the mean of *x* (either in a positive or negative direction), the greater its leverage. High leverage is not synonymous with influence, however. A high leverage observation can follow straight in line with the pattern of the bulk of the data. In this case, the observation is not problematic at all. This will be discussed in more detail shortly.

An observation is *influential* if, when excluded from the analysis, the regression estimates change substantially. The level of influence is determined by the *combination of leverage and discrepancy in terms of y value*. That is, if an observation has high leverage *and* an unusual *y* value given its *x* value, it could strongly *influence* the regression surface. In such cases, both the intercept and slope are affected, as the regression surface chases the observation. Given that OLS estimators are based on the conditional mean of the dependent variable, they suffer from the same problems as the mean itself. Even a single unusual case can influence the coefficient estimates. In other words, OLS estimators have a breakdown point of $BDP(T, Z) = 1/n \cong 0$ (as n gets larger, $1/n$ approaches 0), and its influence function is proportional to the size of the residual. OLS can give distorted estimates if even one problematic case exists.

The various kinds of unusual observations also have different implications for the standard errors of the OLS estimators. Consider the simple regression model, where the standard error for the slope is based on the standard deviation of the residuals s_e and the amount of variation x has around its mean

$$S\hat{E}(\hat{\beta}) = \frac{s_e}{\sqrt{\sum (x_i - \bar{x})^2}}.$$ [3.9]

Large residuals associated with observations in the tails of a heavy-tailed distribution, including outliers, increase the value of s_e, resulting in standard errors that are larger than if the residuals were normally distributed. More specifically, a vertical outlier (i.e., an observation that stands apart from the pattern in the bulk of the data) with low leverage (i.e., its *x* value is not unusual) has no impact on the denominator, but *increases* the numerator and thus the standard error. On the other hand, an observation with high leverage (i.e., an *x* value far from the mean in either a positive or negative direction) increases the size of the denominator, and thus *decreases* the standard error. Simply put, observations that are separated from the rest of the data but follow the general pattern *improve* the precision of the OLS estimators. Only when the observation has an unusual y value given its x value will precision be hindered. In this case, some robust regression estimators have smaller standard errors.

The graphs in Figure 3.1 show how the different types of discrepancy affect the simple regression line. Except for the labeled observations, the data in the three graphs are identical. They were contrived so that there is a strong linear relationship between x_i and y_i when the labeled observation is excluded (see the solid lines). The broken lines show the regressions for the contaminated data (i.e., when the labeled observations are included).

Starting with plot A, observation A has an x value that is not unusual (in fact, it is equal to $\bar{x}$) but a highly unusual y value. In other words, this observation is a regression outlier, but it does not have unusual leverage. As a result, it has no impact on the estimate of the slope. It does have implications for the precision of the estimates, however. Given that it has such a large residual, it will increase the size of the standard errors. This type of observation also pulls the intercept toward it, although the unusualness must be very extreme for it to have a drastic impact (in this case, it has only a small impact). In these situations, a robust regression that down-weights observations according to the size of their residuals (e.g., an M-estimation) would give more precise yet still unbiased estimates.

In graph B, observation B is a univariate outlier in terms of both its y value and its x value but fits perfectly on the regression line. In other words, B has high leverage, but it is not a regression outlier. Although it has no impact on the slope, it decreases the standard error of the estimate because it widens the spread of x. In terms of the OLS estimators, then, this case is not problematic whatsoever. In fact, it would not be sensible to use any method other than OLS.

Finally, in graph C, observation C is unusual both in terms of its x value and its y value given its value of x. In other words, C has high leverage and is a regression outlier, resulting in the regression line being pulled toward it. The greater variability in y due to the outlier also results in a less precisely measured regression coefficient. How to deal with outliers of this nature requires good judgment, and often further research. In the ideal situation, the unusualness results from miscoding, in which case the coding can be fixed. Alternatively, if there is a good rationale for doing so, the observation can be removed and discussed separately as a special case. Another option is to use some form of robust regression—which often gives results nearly equivalent to those associated with removing the discrepant observations. The best choice in this particular situation is a method that considers leverage as well as the residuals (e.g., a GM-estimator, which will

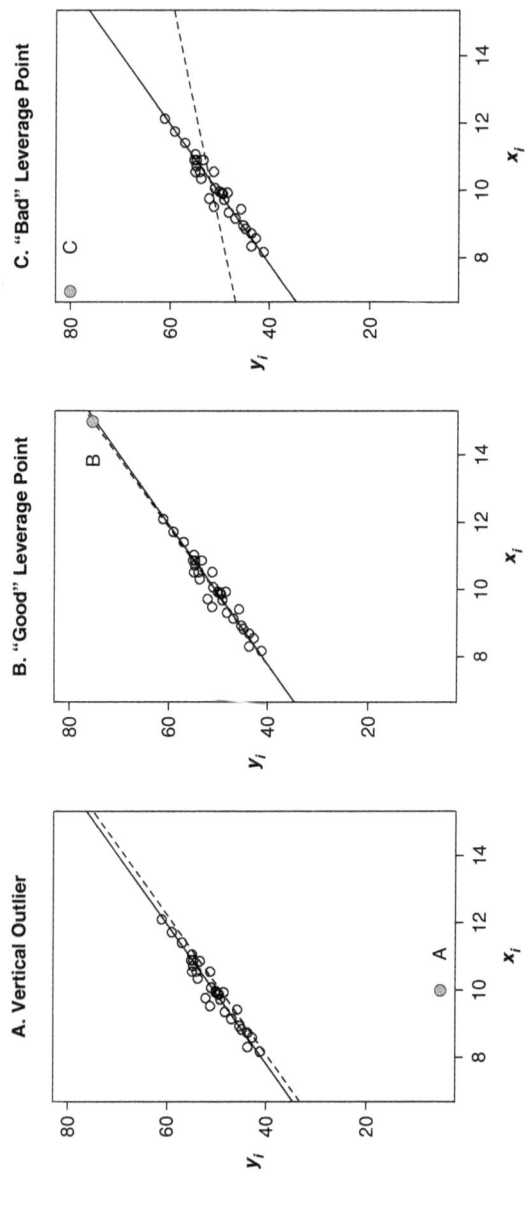

Figure 3.1 Types of Discrepancy and Their Consequences on OLS Estimates

NOTE: The broken line indicates the regression line including the discrepant case; the solid line is the regression line excluding the discrepant case.

be discussed in Chapter 4). As we shall see later, not all robust regression techniques have this attribute.

In summary, a regression outlier with low leverage has little impact on the regression slopes, although it can affect estimates of the intercept. More important, it will negatively affect the fit of the model, and the standard errors of the estimates, because it has a large residual. An observation with high leverage that is not discrepant in terms of its y value—in other words, the observations falls in line with the pattern in the bulk of the data—will also not affect the slope estimates, and its presence will actually improve the fit of the model, making the estimates more precise. Only when discrepancy and leverage are combined are the slope coefficients affected.

This all implies that failure to detect and account for influential cases could lead to distorted conclusions. These conclusions would be based on a poor model regardless of whether or not standard measures for assessing the model—such as the R^2 and the standard errors for coefficients—looked fine. In other words, standard numerical measures of fit cannot always give indications of the impact of discrepant observations on the regression coefficients. This suggests the importance of graphical methods for assessing influence. We must look very carefully at the pattern in the data in order to have confidence in the estimates from the model. Of course, this is a general principle of statistical analysis that does not apply to regression alone.

EXAMPLE 3.1: Income Inequality and Public Opinion Toward Pay Equality in 26 New Democracies

There is no better way to show the limitation of OLS than with an empirical example. Continuing with the public opinion data introduced in the previous chapter, the goal is to predict public opinion from the Gini coefficient. More specifically, we are interested in whether or not the level of income inequality in a country affects public opinion toward pay inequality. The present example concentrates only on the 26 new democracies.

Figure 3.2 shows a scatterplot of the relationship between the two variables. It also includes various regression lines fitted to the data. The solid line predicts public opinion from an OLS regression using information from all 26 countries; the dashed line shows the OLS regression excluding two outliers—the Czech Republic and Slovakia—and the dotted line is from a robust regression using M-estimation, which gives less weight to observations with large residuals in an initial OLS fit. (More details regarding the method are given in Chapter 4.)

The importance of graphical examination of the data is obvious here, as it is easy to see that the OLS regression line is pulled toward the outliers.

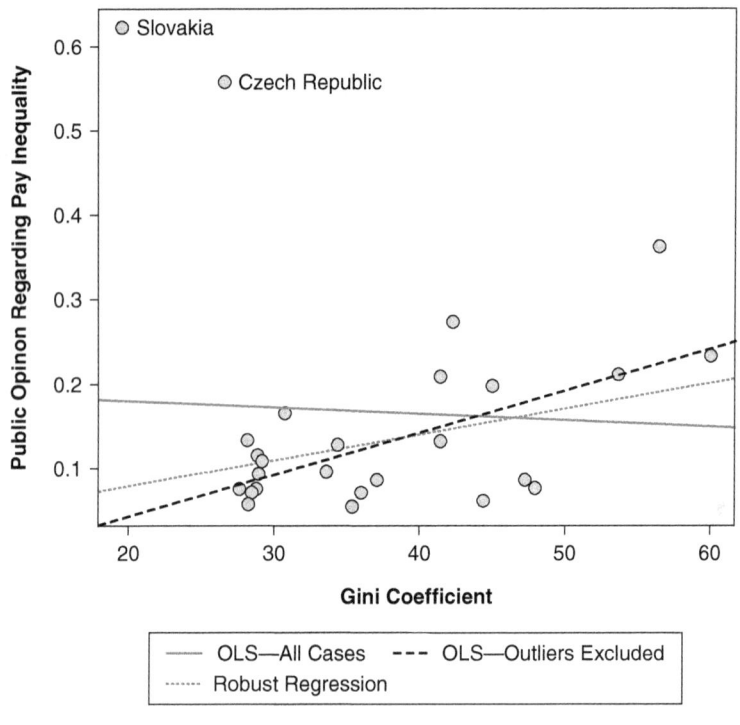

Figure 3.2   Income Inequality and Public Opinion Toward Pay Equality in 26 New Democracies

The two outliers have both high leverage (i.e., their x values, Gini, are a fair distance from the mean $\bar{x}$) and are regression outliers (i.e., they have discrepant values of y, public opinion, given their values of x). In other words, these two cases do not follow the general pattern of the bulk of the data. We also see that the robust regression better reflects the pattern for the majority of cases than does the OLS regression, providing a fit to the data that is somewhat similar to the fit from the OLS regression excluding the outliers.

The numerical output for the regressions is shown in Table 3.1. If we were to report a final regression model before exploring for influential cases, we would perhaps (unwisely) settle on the OLS estimation, concluding that there is no relationship between income inequality and public opinion on pay inequality. In doing so, we would have allowed two observations—the Czech Republic and Slovakia—to entirely influence the result. The OLS regression fitted to the data with these two observations removed indicates a

TABLE 3.1
Regression Models Predicting Public Opinion
on Income Inequality, New Democracies

	OLS (All Observations)		OLS (Omitting Czech Republic and Slovakia)		Robust Regression (M-Estimation)	
	$\hat{\beta}$	SE$(\hat{\beta})$	$\hat{\beta}$	SE$(\hat{\beta})$	$\hat{\beta}$	SE$(\hat{\beta})$
Intercept	0.195	0.111	−0.059	0.053	0.016	0.056
Gini Coefficient	−0.0008	0.0028	0.0050	0.0013	0.0031	0.0015
S_e	0.1485		0.0627			
R^2	0.0029		0.3887			
n	26		24		26	

fairly strong relationship between the two variables. The effect of the Gini coefficient not only becomes positive and statistically significant at conventional levels ($p < .001$), but the regression also has a substantially better fit to the data. Before the outliers are removed, the R^2 was close to 0, but it improves to a respectable 0.39 after they are removed. Likewise, the standard error about the regression line improves substantially after the outliers are removed, falling from about 0.149 to less than half that, at about 0.063. Reflecting what was seen in Figure 3.2, similar to the OLS regression that omits the two outliers, the robust regression estimate for income inequality is positive, although smaller. The effect is also statistically significant ($p = .025$), although because the n is small, the asymptotic standard errors reported in the table should be interpreted cautiously (more on this in Chapter 5).

Detecting Problematic Observations in OLS Regression

The previous chapter showed the importance of detecting and handling unusual observations in simple regression. We now turn to some traditional methods for detecting outliers in multiple regression analysis. For the most part, fairly basic descriptions will be provided. For further details, both of the methods discussed here and of additional methods, see Cook and Weisberg (1982), Rousseeuw and Leroy (1987), Chatterjee and Hadi (1988), and Fox (1991).

We start by fitting two OLS regressions to the cross-national data, both regressing public opinion on the Gini coefficient and per capita GDP. Model 1 includes all 26 new democracies. As Table 3.2 indicates, the effect of the Gini coefficient on public opinion is not statistically significant ($p = .40$), but there is a statistically significant effect of per capita GDP

TABLE 3.2

OLS Regressions Predicting Public Opinion on Income Inequality
From the Gini Coefficient and Per Capita GDP, New Democracies

	OLS (All Observations)		OLS (Omitting Czech Republic and Slovakia)	
	$\hat{\beta}$	SE($\hat{\beta}$)	$\hat{\beta}$	SE($\hat{\beta}$)
Intercept	0.028	0.128	−0.107	0.058
Gini Coefficient	0.00074	0.0028	0.00527	0.0013
Per Capita GDP/1000	0.0175	0.0079	0.0063	0.0037
S_e		0.138		0.0602
R^2		0.175		.4622
n		26		24

($p = .018$). Our previous analyses indicated that two cases—the Czech Republic and Slovakia—had unusually high public opinion scores given their low Gini coefficient. Model 2, which excludes these two observations, leads to very different substantive conclusions. The effect of per capita GDP is no longer statistically significant ($p = .051$) and is only slightly more than a third of its size with the two outliers included. On the other hand, the effect for the Gini coefficient has increased by a factor of more than seven, becoming statistically significant ($p = .0002$). Just as important, the standard error about the regression is more than twice as large, and the R^2 is less than half as large, for the regression that includes the outliers. It is clear, then, that these two observations are causing serious problems for the estimates. Of course, we don't usually have prior knowledge of influential observations until after performing model diagnostics. We now carry out diagnostics for the regression using all 26 observations (i.e., including the two outliers).

Detecting Leverage: Hat Values

We begin by exploring for observations with high leverage (i.e., points with unusual x values). The most common measure of leverage is the hat value h_i, which, as discussed earlier (see Equation 3.7), is the weight that transforms a particular y_i onto its fitted value $\hat{y}_i$. If h_{ij} is large, the ith observation has a substantial impact on the jth fitted value

$$\hat{y}_i = h_{1j}y_1 + h_{2j}y_2 + \cdots + h_{nj}y_n = \sum_{j=1}^{n} h_{ij}y_i. \qquad [3.10]$$

A single hat value h_i measures the potential leverage of a particular y_i observation on *all* the fitted values. Hat values are bounded between $1/n$ and 1, with an average value of $\bar{h} = (k + 1)n$. It is important to note that, in least squares regression, the values of y are irrelevant in the calculation of hat values.[1] Hat values consider only how far a particular value of x_i is from its mean $\bar{x}$. In the simple regression case,

$$h_i = \frac{1}{n} + \frac{(x_i - \bar{x})^2}{\sum_{j=1}^{n} (x_j - \bar{x})^2}. \qquad [3.11]$$

The hat values perform the same function in multiple regression, although rather than measuring distance from the mean of a single x, h_i measures distance from the centroid point of the xs—that is, the meeting point of the means of all of the xs. In other words, hat values in multiple regression take into consideration the correlational and variational structure of the xs (see Cook and Weisberg 1999:161-163). Hat values indicate leverage only—they do not tell us whether the corresponding value of y is unusual given the value of x_i. Although high leverage observations sometimes have large residuals, this is not necessarily so. In fact, precisely because they pull the regression line toward them, high leverage points can have very small residuals,

$$V(e_i) = \sigma_\varepsilon^2 (1 - h_i).$$

Figure 3.3 plots the hat values from Model 1 against the case number of the observations (a so-called index plot of the hat values). Although there is no formal test for leverage, a rule of thumb is that hat values exceeding about twice the average hat value should be considered noteworthy.[2] The dashed line in the plot represents this value. Three observations stand apart from the rest of the data: Brazil, Chile, and Slovenia. Although these observations have high leverage, it is impossible to tell if they stray from the pattern of the bulk of the data without further analysis. Recall that if they are in line with the rest of the data, these are then "good" leverage observations because they help decrease the standard errors of the estimates. So, we make note of these observations but are not yet worried about them, proceeding to explore for regression outliers.

Detecting Regression Outliers: Studentized Residuals and the Bonferroni Adjustment

At first thought, it would seem that the simplest way to detect regression outliers is to calculate standardized residuals e_i', and identify observations with $|e_i'| > 2$ as significant outliers (i.e., residuals that are larger than two

40

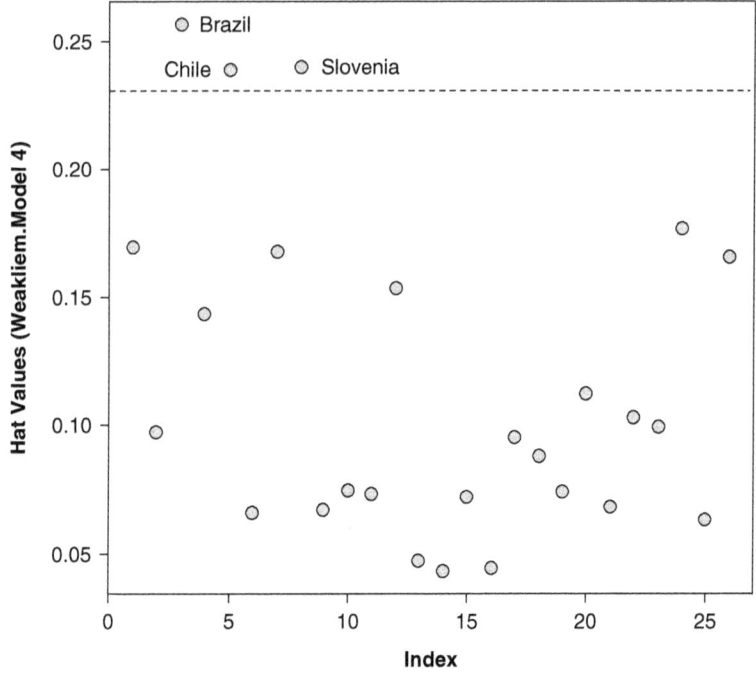

Figure 3.3 Hat Values From Model 1

standard deviations from the mean of the residuals). These values are problematic for inference, however, because the numerator and the denominator of the equation are not independent, and thus do not follow a t distribution. As we see below, the residual of interest e_i is not just on the numerator, but is also included in the calculation of the standard deviation of the residuals s_e, which appears in the denominator:

$$e'_i = \frac{e_i}{s_e\sqrt{1 - h_i}}. \qquad [3.12]$$

The solution is to calculate *studentized residuals*, e'_i, which remove the observation of interest from the calculation of the standard deviation of the residuals

$$e^*_i = \frac{e_i}{s_{e(-i)}\sqrt{1 - h_i}}. \qquad [3.13]$$

This results in a t distribution with $n - k - 2$ degrees of freedom.

A quantile comparison plot that compares the studentized residuals with quantities of the theoretical t distribution of the same degrees of freedom is very useful for detecting outliers, especially, as is commonly the case, when we don't suspect any particular observations to be discrepant before fitting the regression model. Figure 3.4 is a quantile comparison plot of the residuals from Model 1. The 95% confidence envelope surrounding the observations was constructed using bootstrapping, a topic to be covered in Chapter 5. Two observations stray from the confidence envelope: the Czech Republic and Slovakia. Recall that these cases were not identified as having extreme hat values.

We can also formally test whether a particular observation is an outlier, although the standard p values can't be trusted. If we intentionally test the most extreme residual—which is common practice—rather than randomly select an observation, the test is biased toward statistical significance because even if the residuals are normally distributed, we should expect 5% of the studentized residuals to be statistically significant by chance alone (at an alpha level of .05). A Bonferroni adjustment to the p value for a two-sided t test for the largest outlier remedies this problem. The Bonferroni p value $= np'$ where p' is the unadjusted p value from a t test with $n-k-2$ degrees of freedom (Fox 1997:274). The t statistic for Slovakia—the observation with the largest residual—is 4.31 with 22 degrees of freedom and a p value $= .00027$. With the Bonferroni adjustment, the p value is $26 \times .00027 = .0072$, indicating that Slovakia is a significant outlier. Although we have identified discrepant cases, we must remember that unusual observations do not always have large residuals.

Detecting Influence: DFBETAs, Cook's D, and Partial Regression Plots

The most direct approach to assessing influence is to examine how the regression coefficients change if an outlier is omitted. We can use what Belsley, Kuhn, and Welsch (1980) call DFBETAs (or D_{ij}), which are defined quite simply by

$$D_{ij} = \hat{\beta}_j - \hat{\beta}_{j(-i)}, \text{ for } i = 1, \ldots, n \text{ and } j = 0,1,\ldots,k,$$

where the β_j are for all of the data and the $\hat{\beta}_{j(-i)}$ are with the ith observation removed. Each observation has a separate DFBETA$_{ij}$ for each coefficient. There is no formal test for significance for the DFBETA$_{ij}$, but a common rule of thumb cutoff is $|D_{ij}| \geq 2/\sqrt{n}$. It can be helpful to plot the DFBETA$_{ij}$ against the case numbers, looking for values that are large relative to others.

Returning to Model 1, index plots for the DFBETA$_i$ for per capita GDP and for the Gini coefficient are shown in Figure 3.5. The dotted

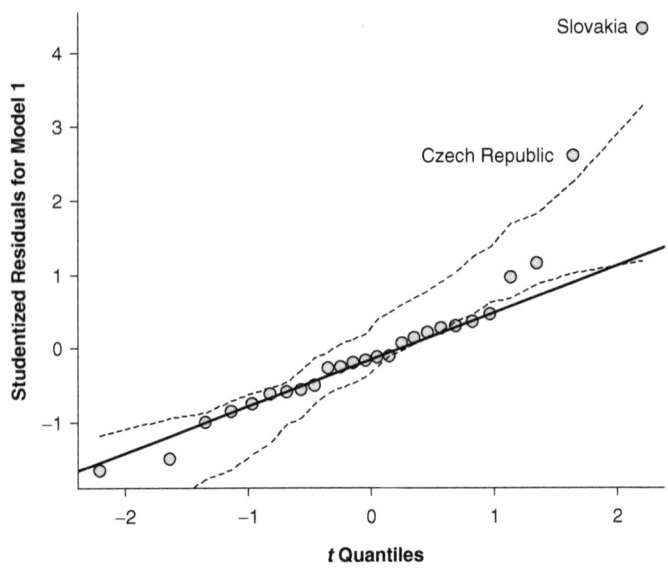

Figure 3.4   Quantile Comparison Plot for Studentized Residuals From Model 1

lines represent the rule of thumb cutoffs, $|D_{ij}| \geq 2/\sqrt{n}$. We already discovered that the Czech Republic and Slovakia were regression outliers. We now see that they also have unusually high influence on the regression estimates. This influence is with respect to both predictors. Slovakia substantially pulls the coefficient for per capita GDP in a positive direction and the effect of the Gini coefficient in a negative direction. Although it has a stronger impact on the per capita GDP effect and a weaker impact on the Gini coefficient than does Slovakia, the general pattern is similar for the Czech Republic. Three other potentially problematic cases are also identified: Chile pulls the Gini coefficient in a positive direction; Taiwan and Slovenia pull the per capita GDP coefficient in a negative direction.

DFBETAs are useful for understanding on which predictor an observation has influence, but because we require a separate measure for each observation on each coefficient, they become cumbersome as dimensionality increases. Cook's Distances, also known as Cook's D, overcomes this problem by providing a single measure of overall influence on the regression surface for each observation (Cook 1977). Other methods related to the Cook's D are the DFFITs (Belsley et al. 1980) and Atkinson's Modified

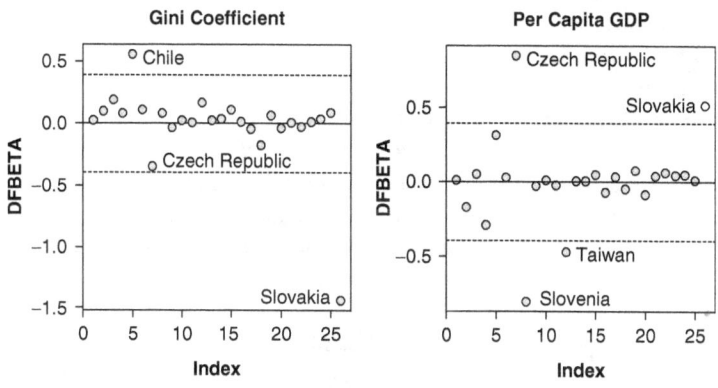

Figure 3.5   Index Plots of the DFBETAs for the Coefficients for Per Capita
GDP and Gini Coefficient From Model 1

Cook's Statistic (Atkinson 1985). These methods generally give similar
results (see Draper and Smith 1998:214), so we focus only on the more
commonly employed Cook's D.

The Cook's D for a particular observation is

$$D_i = \frac{e_i'^2}{k+1} \times \frac{h_i}{1 - h_i},$$ [3.14]

where k is the number of predictors. The first term of the equation contains
the standardized residual for the observation, e_i', and hence measures
discrepancy. The second term includes the hat value, h_i, and thus measures
leverage. Although there is no formal test of statistical significance for
Cook's D, rough cutoffs have been proposed. Cook and Weisberg
(1999:358) suggest closely exploring observations with $D_i > 0.5$. Based on
Chatterjee and Hadi's (1988) cutoff for the related DFFITs measure, Fox
(1997:281) suggests a cutoff that is dependent on sample size, n, and the
number of predictors, k, $D_i > \frac{4}{n-k-1}$. These cutoffs are useful, but they are
not always successful at distinguishing problematic cases, and thus there is
no substitution for examining relative discrepancies.

An index plot of Cook's Ds helps identify the relative overall influence
of observations on the regression estimates (Cook and Weisberg 1999:358).
Another option is to construct an "influence plot," which is attributed to
Fox (1991:37-38).[3] The utility of the influence plot is that it shows the relative weight of discrepancy and leverage in determining influence. The influence plot is constructed by plotting the studentized residuals, e_i^*, on the

vertical axis against the hat values h_i on the horizontal axis. Each observation is then represented by a hollow circle with an area proportional to Cook's D. In other words, the larger the circle representing the observation, the greater the influence the observation has on the regression surface. Figure 3.6 displays an index plot of Cook's D and an influence plot for Model 1. Unusual observations are labeled in both plots. The high influence of Slovakia and the Czech Republic is evident by the relatively large circles representing them in the influence plot. We also see quite clearly that, for both observations, the influence results from a combination of discrepancy and leverage. We should also note that Slovenia also has much higher influence on the regression estimates than other countries.

Cook's D can be very successful at identifying influential cases when there are relatively few of them, but it can fail to detect jointly influential observations, especially if there are several of them. In such cases, it is possible for none of the observations to have highly unusual influence on its own, and hence it does not have a large Cook's D. If there are very few jointly influential observations, Cook's D can be used to identify them by sequentially deleting influential observations and continually updating the model, exploring the Cook's Ds each time. But this approach is impractical if there are a large number of subsets to explore. Partial regression plots, also called added variable plots, are a much more useful method, at least for identifying observations that have influence on a single coefficient (Cook and Weisberg 1999:360).[4]

A partial regression plot is similar to a simple scatterplot in that all observations are plotted on the graph. It differs, however, in that the pattern in the observations indicates the *partial relationship* between y and x rather than the *marginal relationship*. In other words, partial regression plots graph the effect of a predictor while holding all other predictors constant. Let $y_i^{(1)}$ represent the residuals from the least squares regression of y on all of the xs except for x_1, $y_i = a^{(1)} + \hat{\beta}_2^{(1)} x_{i2} + \cdots + \hat{\beta}_k^{(1)} x_{ik} + y_i^{(1)}$. Similarly, let $x_i^{(1)}$ represent the residuals from the regression of x_1 on all other xs (but *not* y), $x_{i1} = c^{(1)} + d_2^{(1)} x_{i2} + \cdots + d_k^{(1)} x_{ik} + x_i^{(1)}$. These two equations determine the residuals $y_i^{(1)}$ and $x_i^{(1)}$ as parts of y and x_1 that remain after the linear effects of $x_2, \ldots, x_k$ have been removed. The residuals $y_i^{(1)}$ and $x_i^{(1)}$ have three convenient properties for assessing leverage: (1) the slope of the regression of $y_i^{(1)}$ on $x_i^{(1)}$ is the least squares slope $\hat{\beta}_1$ from the multiple regression (i.e., it is the same as the partial regression slope); (2) the residuals from the regression of $y_i^{(1)}$ on $x_i^{(1)}$ are identical to the residuals from the original multiple regression,

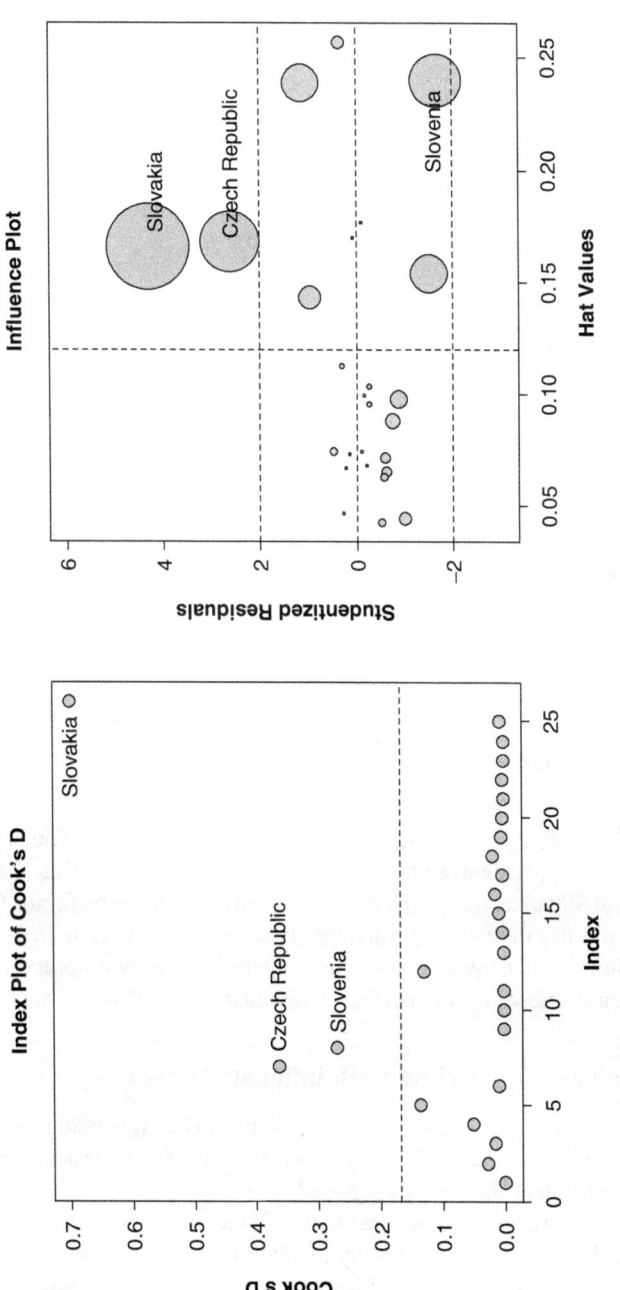

Figure 3.6 Assessing Influence in Model 1

45

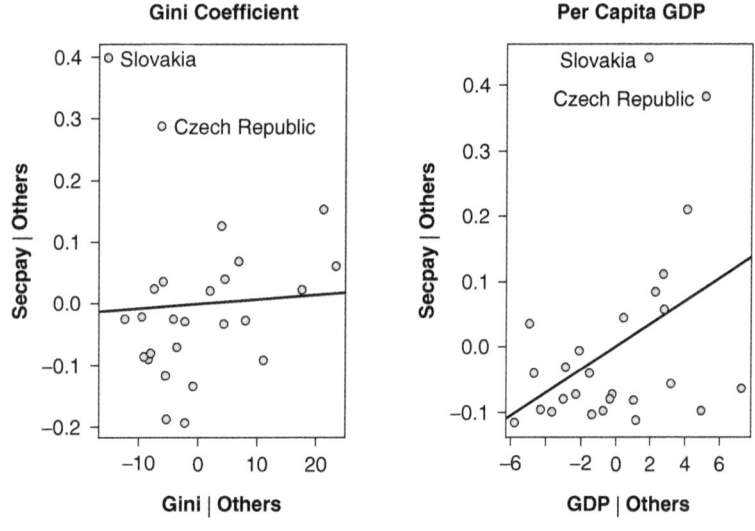

Figure 3.7   Partial Regression Plots for Model 1

$y_i^{(1)} = \hat{\beta}_1 x_1^{(1)} + e_i$; and (3) variation of $x^{(1)}$ is the conditional variance of x_1 holding the other xs constant. Because of these attributes, a plot of $y_i^{(1)}$ against $x_i^{(1)}$ can indicate both individual and joint influence on $\hat{\beta}_1$.

Partial regression plots for Model 1 are shown in Figure 3.7. The partial regression plots provide even more evidence of the influence of the Czech Republic and Slovakia, indicating that they jointly pull the slope for the Gini coefficient in a negative direction and the slope for per capita GDP in a positive direction. In other words, combined together, these two observations have an even stronger impact than their individual Cook's D would suggest.

Some Strategies for Dealing With Influential Cases

Once problematic cases have been detected in an OLS regression, several options can be considered: (1) Investigate whether the deviations are a symptom of model failure that can be repaired by recoding or removing cases; (2) transform a variable (or variables) to rectify the problem, especially with respect to a skew; (3) if there are many jointly influential cases, we could try adding more terms to the model—either new variables, or allow existing predictors to interact—to account for the unusual pattern exhibited by these observations; or (4) use a method that is more robust and resistant to outliers. We now turn our attention to various robust regression techniques.

Notes

1. As we shall see in Chapter 6, this is not the case for generalized linear models.

2. With large sample sizes, however, this cutoff is unlikely to identify any observations regardless of whether they deserve attention (Fox 1991).

3. Fox's Influence Plot can be routinely implemented using the influence.plot function in the car package for **R** (see Fox 2002:198 for more details).

4. Partial regression plots should not be confused with the similar partial residual plots. The latter typically are not as effective for assessing influence but tend to be better at distinguishing between monotone and nonmonotone nonlinearity. For more details on the relative merits of the partial regression plots and partial residual plots, see Fox (1997).

4. ROBUST REGRESSION FOR THE LINEAR MODEL

We now explore various robust regression techniques—including those sometimes labeled as resistant regression techniques—in an evolutionary manner, explaining how new methods evolved in response to limitations of existing ones. Several classes of regression will be discussed: L-estimators (based on linear combinations of order statistics); R-estimators (based on the ranks of the residuals); M-estimators (extending from M-estimates of location by considering the size of the residuals); GM-estimators (or generalized M-estimators, which extend M-estimators by giving less weight to high influence points as well as to large residual points); S-estimators (which minimize a robust M-estimate of the residual scale); and MM-estimators (which build on both M-estimation and S-estimation to achieve a high breakdown point with high asymptotic efficiency). Some of these methods should be considered obsolete, but general descriptions are still provided because more recent developments in robust regression build on them. The chapter ends with a discussion of how robust regression can be used as a diagnostic method for identifying problematic cases.

L-Estimators

Any estimator that is computed from a linear combination of order statistics can be classified as an L-estimator. The first L-estimation procedure, which is somewhat more resistant than OLS, is least absolute values (LAV) regression. Also known as L_1 regression[1] because it minimizes the L_1-norm (i.e., sum of absolute deviations), LAV is the simplest and earliest approach to bounded influence robust regression, predating OLS by about 50 years (Wilcox 2005:451). Least squares regression also fits this definition, and thus it is sometimes referred to as L_2, reflecting that the L_2-norm (i.e., the sum of squared deviations) is minimized. Other well-known L-estimators are the least median of squares and the least trimmed squares estimators.[2]

Least Absolute Values Regression

Least absolute values (LAV) regression is very resistant to observations with unusual y values. Estimates are found by minimizing the sum of the *absolute values of the residuals*

$$\min \sum_{i=1}^{n} |e_i| = \min \sum_{i=1}^{n} \left| y_i - \sum x_{ij}\beta_j \right|. \qquad [4.1]$$

The LAV can be seen as a case of the more general quantile regression. In this case, the objective function to be minimized can be written as

$$\sum_{i=1}^{n} \rho_\alpha(e_i), \qquad [4.2]$$

where

$$\rho_\alpha(e_i) = \begin{cases} \alpha e_i & \text{if } e_i \geq 0 \\ (\alpha - 1)e_i & \text{if } e_i < 0 \end{cases} \qquad [4.3]$$

and α is the quantile being estimated. For general applications of quantile regression, see Koenker and Bassett (1978; see also Koenker and d'Orey 1994; Koenker 2005). For a treatment geared toward social scientists, see Hao and Naiman (2007).[3]

Although LAV is less affected than OLS by unusual y values, it fails to account for leverage (Mosteller and Tukey 1977:366), and thus has a break-down point of $BDP = 0$. Moreover, LAV estimates have relatively low efficiency. Following the case of the mean, under the assumption that $y \sim N(\mu, \sigma^2)$, the sampling variance of y for OLS is σ^2/n; for LAV it is $\pi/2 = 1.57$ times larger at $\pi\sigma^2/2n$ (in other words, about 64% efficiency). The combination of the low breakdown point and low efficiency makes LAV less attractive than other robust regression methods still to be discussed.

Least Median of Squares Regression

First proposed by Rousseeuw (1984), least median of squares (LMS)[4] replaces the summing of the squared residuals that characterizes OLS with the median of the squared residuals. The estimates are found by

$$\min M \left(y_i - \sum x_{ij}\beta_j \right)^2 = \min M \left(e_i^2 \right),$$ [4.4]

where M denotes the median. The idea is that by replacing the sum with the more robust median, the resulting estimator will be resistant to outliers. Although this result is achieved (it has a breakdown point of $BDP = 0.5$), the LMS estimator has important deficiencies that limit its use. It has at best a relative efficiency of 37% (see Rousseeuw and Croux 1993), and it does not have a well-defined influence function because of its convergence rate of $n^{-1/3}$ (Rousseeuw 1984). Despite these limitations, as we shall see later, LMS estimates can play an important role in the calculation of the much more efficient MM-estimators by providing initial estimates of the residuals.

Least Trimmed Squares Regression

Another method developed by Rousseeuw (1984) is least trimmed squares (LTS) regression. Extending from the trimmed mean, LTS regression minimizes the sum of the trimmed squared residuals. The LTS estimator is found by

$$\min \sum_{i=1}^{q} e_{(i)}^2,$$ [4.5]

where $q = [n(1 - \alpha) + 1]$ is the number of observations included in the calculation of the estimator, and α is the proportion of trimming that is performed. Using $q = (n/2) + 1$ ensures that the estimator has a breakdown point of $BDP = 0.5$. Although highly resistant, LTS suffers badly in terms of relative efficiency at about 8% (see Stromberg, Hossjer, and Hawkins 2000). Its efficiency is so low that it is not desirable as a stand-alone estimator. Still, the LTS has merit in the role it plays in the calculation of other estimators. For example, the GM-estimators proposed by Coakley and Hettmansperger (1993) use LTS to obtain initial estimates of the residuals. LTS residuals can also be used effectively in outlier diagnostic plots, to be discussed later.

R-Estimators

First proposed by Jaeckel (1972), R-estimators rely on dispersion measures that are based on the linear combinations of the ordered residuals (i.e., on the rank of the residuals). Let R_i represent the rank of the residuals e_i. R-estimators minimize the sum of some score of the ranked residuals

$$\min \sum_{1-1}^{n} a_n(R_i)e_i \qquad [4.6]$$

where $a_n(i)$ is a monotone score function that satisfies

$$\sum_{i=1}^{n} a_n(i) = 0. \qquad [4.7]$$

Many possibilities have been proposed for the score function. The simplest, and perhaps most commonly employed, are the Wilcoxon Scores, which directly find the rank of observations from the median

$$a_n(i) = i - \left(\frac{n+1}{2}\right). \qquad [4.8]$$

Median Scores are a simple adjustment over the Wilcoxon scores,

$$a_n(i) = \sin\left[i - \left(\frac{n+1}{2}\right)\right] \qquad [4.9]$$

Van der Waerden Scores adjust the ranks according to the inverse of the normal probability density function Φ^{-1}:

$$a_n(i) = \Phi^{-1}\left(\frac{i}{n+1}\right) \qquad [4.10]$$

Finally, Bounded Normal Scores adjust the Van der Waerden Scores by bounding them according to a constant, c:

$$a_n(i) = \min\left\{c, \max\left[\Phi^{-1}\left(\frac{i}{n+1}\right), -c\right]\right\} \qquad [4.11]$$

An advantage of R-estimators over some others (such as M-estimators, and those extending from them) is that they are scale equivariant. They have some undesirable attributes, however. One problem is that the optimal choice for the score function is unclear. A second problem is that the objective function is invariant with respect to the intercept. If an intercept is not required, this is of no concern—it is simply not estimated. Even if one is needed, it can be calculated manually after fitting the model from the median of the residuals, so this limitation is surmountable. More problematic is the fact that most R-estimators have a breakdown point of $BDP = 0$. An exception is the bounded influence R-estimator of Naranjo and Hettmensperger (1994), which is also fairly efficient (90%–95%) when the Gauss-Markov assumptions are met. Even for this estimator, however, the breakdown point never reaches more than 0.20. As a result, we leave R-estimates behind, proceeding to more robust estimators. (For more extensive details of R-estimates, see Huber 2004; Davis and McKean 1993; McKean and Vidmar 1994.)

M-Estimators

First proposed by Huber (1964, 1973, 2004), M-estimation for regression is a relatively straightforward extension of M-estimation for location. It represents one of the first attempts at a compromise between the efficiency of the least squares estimators and the resistance of the LAV estimators, both of which can be seen as special cases of M-estimation. In simple terms, the M-estimator minimizes some function of the residuals. As in the case of M-estimation of location, the robustness of the estimator is determined by the choice of weight function.

If we assume linearity, homoscedasticity, and uncorrelated errors, the maximum likelihood estimator of β is simply the OLS estimator found by minimizing the sum of squares function

$$\min \sum_{i=1}^{n} \left(y_i - \sum x_{ij}\beta_j \right)^2 = \min \sum_{i=1}^{n} (e_i)^2. \qquad [4.12]$$

Following from M-estimation of location, instead of minimizing the sum of squared residuals, a robust regression M-estimator minimizes the sum of a less rapidly increasing function of the residuals

$$\min \sum_{i=1}^{n} \rho \left(y_i - \sum x_{ij}\beta_j \right) = \min \sum_{i=1}^{n} \rho(e_i). \qquad [4.13]$$

The solution is not scale equivariant, and thus the residuals must be standardized by a robust estimate of their scale $\hat{\sigma}_e$, which is estimated simultaneously. As in the case of M-estimates of location, the median absolute deviation (MAD) is often used. Taking the derivative of Equation 4.13 and solving produces the score function

$$\sum_{i=1}^{n} \Psi \left(y_i - \sum x_{ij}\beta_j \Big/ \hat{\sigma} \right) x_{ik} = \sum_{i=1}^{n} \Psi(e_i/\hat{\sigma}_e) \mathbf{x}_i = 0 \qquad [4.14]$$

with $\Psi = \rho'$. There is now a system of $k+1$ equations, for which Ψ is replaced by appropriate weights that decrease as the size of the residual increases

$$\sum_{i=1}^{n} w_i(e_i/\hat{\sigma}_e)\mathbf{x}_i = 0. \qquad [4.15]$$

Iteratively Reweighted Least Squares

An iterative procedure is necessary to find M-estimates for regression. A single step is impossible because the residuals can't be found until the model is fitted, and the estimates can't be found without knowing the residuals. As a result, iteratively reweighted least squares (IRLS) is employed[5]:

1. Setting the iteration counter at $I = 0$, an OLS regression is fitted to the data, finding initial estimates of the regression coefficients $\hat{\beta}^{(o)}$.
2. The residuals are extracted from the preliminary OLS regression, $e_i^{(0)}$, and used to calculate initial estimates for the weights.
3. A weight function is then chosen and applied to the initial OLS residuals to create *preliminary* weights, $w(e_i^{(0)})$.
4. The first iteration, $I = 1$, uses weighted least squares (WLS) to minimize $\sum w_i^{(1)} e_i^2$ and thus obtain $\hat{\beta}^{(1)}$. In matrix form, with $\mathbf{W}$ representing the $n \times n$ diagonal matrix of individual weights, the solution is

$$\hat{\beta}^{(1)} = \left(\mathbf{X}^T \mathbf{W} \mathbf{X}\right)^{-1} \mathbf{X}^T \mathbf{W} \mathbf{y}. \qquad [4.16]$$

5. The process continues by using the residuals from the initial WLS to calculate new weights, $w_i^{(2)}$.
6. The new weights $w_i^{(2)}$ are used in the next iteration, $I = 2$, of WLS to estimate $\hat{\beta}^{(2)}$.
7. Steps 4–6 are repeated until the estimate of $\hat{\beta}$ stabilizes from the previous iteration.

More generally, at each of the q iterations, the solution is $\hat{\beta}^{(I)} = \left(\mathbf{X}^T \mathbf{W}_q \mathbf{X}\right)^{-1} \mathbf{X}^T \mathbf{W}_q \mathbf{y}$, where $\mathbf{W}_q_{(n \times n)} = \text{diag}\left\{w_i^{(I-1)}\right\}$. The iteration process continues until $\hat{\beta}^{(I)} - \hat{\beta}^{(I-1)} \cong 0$. Typically, the solution is considered to have converged when the change in estimates is no more than 0.01% from the previous iteration. We return to IRLS in more detail in Chapter 6 with respect to robust generalized linear models.

M-estimators are defined to be robust against heavy-tailed error distributions and nonconstant error variance—and thus y outliers—but they also implicitly assume that the model matrix $\mathbf{X}$ is measured without error. Under these conditions, M-estimates are more efficient than OLS estimates. Under the Gauss-Markov assumptions, however, M-estimates are about 95% as efficient as OLS estimates.[6] Moreover, although M-estimators are an improvement over OLS in terms of resistance and robustness to regression outliers (i.e., unusual y values given their xs), like LAV estimators, they are not completely immune to unusual observations because they do not consider leverage. Recall that M-estimates of location are highly robust, having a bounded influence function and a breakdown point of $BDP = 0.5$. M-estimates for regression share these attributes for y but not for the xs, resulting in a breakdown point of $BDP = 0$. In other words, in some situations they perform no better than OLS (see Rousseeuw and Leroy 1987). As we shall see later, these estimators are still important because of the role they play in computing other, more robust estimates. Because these newer estimates perform much better, they should generally be preferred over the original M-estimation.

GM-Estimators

The M-estimator has unbounded influence because it fails to account for leverage (Hampel et al. 1986). In response to this problem, bounded influence Generalized M-estimators (GM-estimators) have been proposed. The goal was to create weights that consider both vertical outliers *and* leverage. Outliers are dealt with using a standard M-estimator, and leverage points are typically down-weighted according to their hat value. The general GM class of estimators is defined by

$$\sum_{i=1}^{n} w_i(\mathbf{x}_i)\Psi\left\{\frac{e_i}{v(\mathbf{x}_i)\hat{\sigma}_e}\right\}\mathbf{x}_i = 0, \qquad [4.17]$$

where Ψ is the score function (as in the case of M-estimation, this is typically the Huber or biweight function), and the weights w_i and v_i initially depend on the model matrix $\mathbf{X}$ from an initial OLS regression fitted to the data but are updated iteratively.

The first GM-estimator proposed by Mallows (see Krasker and Welsch 1982) includes only the w_i weights—that is, $v_i(\mathbf{x}_i) = 1$ in Equation 4.17. The w_i are calculated from the hat values. Because hat values range from 0 to 1, a weight of $w_i = \sqrt{1 - h_i}$ ensures that observations with high leverage receive less weight than observations with small leverage (i.e., if $h_i > h_j$, $u_i < u_j$). Although this strategy seems sensible at first, it is problematic because even "good" leverage points that fall in line with the pattern in the bulk of the data are down-weighted, resulting in a loss of efficiency.

Schweppe's solution (introduced in Handschin et al. 1975) adjusts the leverage weights according to the size of the residual e_i. In order to achieve this result, the w_i weights are defined in the same way as for Mallows, $w_i = \sqrt{1 - h_i}$, but now $v_i(\mathbf{x}_i) = w_i$ (see Chave and Thomson 2003). Although the breakdown point for Schweppe's estimators is better than for regular M-estimators that don't consider leverage, Maronna, Butos, and Yohai (1979) show that it is never higher than $1/(p + 1)$, where p is the number of parameters estimated by the model. In other words, as dimensionality increases, the breakdown point gets closer to $BDP = 0$. This is especially problematic because as the number of variables in the model increases, detection of influential cases also becomes increasingly more difficult. Moreover, because they down-weight according to x values without considering how the corresponding y values fit with the pattern of the bulk of the data, efficiency is still hindered (see Krasker and Welsch 1982). Other evidence also suggests that the Schweppe estimator is not consistent when the errors are asymmetric (Carroll and Welsh 1988), meaning that they are ineffective for the more common problem of outliers in one of the tails, the main concern of the present book.

In an attempt to overcome these problems, other GM-estimation procedures completely remove severe outliers and then use M-estimation on the remaining "good" observations (Coakley and Hettmansperger 1993; Chave and Thomson 2003). Perhaps the most notable of these is Coakley and Hettmansperger's (1993) *Schweppe one-step estimator* (S1S), which extends from the original Schweppe estimator. The advantage of this estimator over the original is that the leverage weights consider where observations fit with the bulk of the data. In other words, it considers whether the observations are "good" or "bad" leverage points, giving less weight to the latter. This results in 95% efficiency relative to OLS estimators under the Gauss-Markov assumptions.

The S1S estimator takes initial estimates of the residuals and the scale of the residuals from a regression with a high breakdown point rather than from an OLS regression, as is the case with the GM-estimators developed before it. Using Rousseeuw's LTS estimator for the initial estimates gives a breakdown point of $BDP = 0.5$. The method is also different from the Mallows and Schweppe estimations in that once the initial estimates from the LTS regression are included, final M-estimates are calculated in a single step (hence the name "one-step") rather than iteratively. Although S1S estimators are more efficient than other GS-estimators, and in fact quite comparable to OLS estimators under normality and large sample sizes, simulation studies suggest that their efficiency is very low when n is small (see Wilcox 2005:438-440).

S-Estimators

In response to the low breakdown point of M-estimators, Hampel (1975) suggested considering the scale of the residuals. Following this idea, Rousseeuw and Yohai (1984; see also Rousseeuw and Leroy 1987) proposed S-estimates. S-estimates are the solution that finds the smallest possible dispersion of the residuals

$$\min \hat{\sigma}\left(e_1(\hat{\beta}), \ldots e_n(\hat{\beta})\right). \qquad [4.18]$$

The parallel with OLS, which minimizes the variance of the residuals, should be obvious. Hence, OLS can be seen as a special, less robust case of S-estimation. Rather than minimize the variance of the residuals, robust S-estimation minimizes a robust M-estimate of the residual scale

$$\frac{1}{n}\sum_{i=1}^{n} \rho\left(\frac{e_i}{\hat{\sigma}_e}\right) = b, \qquad [4.19]$$

where b is a constant defined as $b = E_\Phi[\rho(e)]$ and Φ represents the standard normal distribution. Differentiating Equation 4.19 and solving results in

$$\frac{1}{n}\sum_{i=1}^{n} \Psi\left(\frac{e_i}{\hat{\sigma}_e}\right) = b, \qquad [4.20]$$

where Ψ is replaced with an appropriate weight function. As with most M-estimation procedures, either the Huber weight function or the biweight function is usually employed. Although S-estimates have a breakdown point of $BDP = 0.5$, it comes at the cost of very low efficiency (approximately 30%) relative to OLS (Croux, Rousseeuw, and Hossjer 1994).

Generalized S-Estimators

Croux et al. (1994) propose *Generalized S-estimates* (*GS*-estimates) in an attempt to overcome the low efficiency of the original S-estimators. These estimators are computed by finding a *GM*-estimator of the scale of the residuals. A special case of the *GS*-estimator is the *least quartile difference estimator* (LQD), the parallel of which is using the interquartile range to estimate the scale of a variable. The LQD estimator is defined by

$$\min Q_n(e_1, \ldots, e_n), \qquad [4.21]$$

where

$$Q_n = \{|e_i - e_j|; i < j\} \binom{h_p}{2} \cdot \binom{n}{2} \qquad [4.22]$$

and

$$h_p = \frac{n+p+1}{2} \qquad [4.23]$$

and p is the number of parameters in the model. Put more simply, this means that Q_n is the $\binom{h_p}{2}$th order statistic among the $\binom{n}{2}$ elements of the set $\{|e_i - e_j|; i < j|\}$. Although these estimators are more efficient than *S*-estimators, they have a "slightly increased worst-case bias" (Croux et al. 1994:1271).

Yohai and Zamar's (1988) τ estimates are also defined by the minimization of an estimate for the scale of the residuals, but the weights are adaptive depending on the underlying error distribution. This results in a high breakdown point and a high efficiency estimate of the scale of the errors. Nevertheless, points with high leverage are not considered, so the estimator's efficiency is still hindered. Ferretti et al. (1999) tried to overcome this limitation with *generalized τ estimates*, which use weights that consider observations with high leverage in much the same way as *GM*-estimates extend from *M*-estimates. The method achieves a high breakdown point (as high as 0.5) and higher efficiency (though still only about 75%) than other *GS*-estimates. Still, 75% efficiency is low compared to many other estimators, limiting the use of *S*-estimators as stand-alone estimators. On the other hand, because they are highly resistant to outliers, *S*-estimators play an important role in calculating *MM*-estimates, which are far more efficient.

MM-Estimators

First proposed by Yohai (1987), *MM*-estimators have become increasingly popular and are perhaps now the most commonly employed robust regression technique. They combine a high breakdown point (50%) with good efficiency (approximately 95% relative to OLS under the Gauss-Markov assumptions). The "*MM*" in the name refers to the fact that more than one *M*-estimation procedure is used to calculate the final estimates. Following from the *M*-estimation case, iteratively reweighted least squares (IRLS) is employed to find estimates. The procedure is as follows:

1. Initial estimates of the coefficients $\hat{\beta}^{(1)}$ and corresponding residuals $e_i^{(1)}$ are taken from a highly resistant regression (i.e., a regression with a breakdown point of 50%). Although the estimator must be consistent, it is not necessary that it be efficient. As a result, *S*-estimation with Huber or bisquare weights (which can be seen as a form of *M*-estimation) is typically employed at this stage.[7]

2. The residuals $e_i^{(1)}$ from the initial estimation at Stage 1 are used to compute an *M*-estimation of the scale of the residuals, $\hat{\sigma}_e$.

3. The initial estimates of the residuals $e_i^{(1)}$ from Stage 1 and of the residual scale $\hat{\sigma}_e$ from Stage 2 are used in the first iteration of weighted least squares to determine the *M*-estimates of the regression coefficients

$$\sum_{i=1}^{n} w_i \left(e_i^{(1)} \Big/ \hat{\sigma}_e \right) \mathbf{x}_i = 0, \qquad [4.24]$$

where the w_i are typically Huber or bisquare weights.

4. New weights are calculated, $w_i^{(2)}$, using the residuals from the initial WLS (Step 3).

5. Keeping constant the measure of the scale of the residuals from Step 2, Steps 3 and 4 are continually reiterated until convergence.

Comparing the Various Estimators

Table 4.1 summarizes some of the robustness attributes of most of the estimators we have discussed. Reported are the breakdown point, whether or not the estimator has a bounded influence function, and the approximate asymptotic efficiency of the estimator relative to the OLS estimator. Immediately obvious is the comparatively low breakdown point for the LAV and *M*-estimators, which, depending on how the data are configured, can sometimes perform no better than OLS estimators. A single discrepant observation can render these estimates useless. The bounded influence *R*-estimators don't do much better, having a breakdown point of less than $BDP = 0.2$. These methods, at least on their own, should be ignored in favor of others.

TABLE 4.1
Robustness Attributes of Various Regression Estimators

Estimator	Breakdown Point	Bounded Influence	Asymptotic Efficiency
OLS	0	No	100
LAV	0	Yes	64
LMS	.5	Yes	37
LTS	.5	Yes	8
LTM	.5	Yes	66
Bounded R-estimates	< .2	Yes	90
M-estimates (Huber, biweight)	0	No	95
GM-estimates (Mallows, Schweppe)	$1/(p+1)$	Yes	95
GM-estimates (S1S)	.5	Yes	95
S-estimates	.5	Yes	33
GS-estimates	.5	Yes	67
Generalized estimates	.5	Yes	75
MM-estimates	.5	Yes	95

We should also be cautious of estimators with low efficiency, such as the LMS, LTS, LTM, and S-estimates. If the goal is only to ensure resistance, and not to make inferences about a population, then these estimates may be appropriate. On the other hand, one should not use them without good knowledge of the nature of the unusual observations. Using them blindly could result in less efficient estimates than otherwise possible. For example, if the errors are normally distributed, it would be better to use OLS estimators.

Despite its low breakdown point, the efficiency of the M-estimates is a favorable attribute. When used in combination with more resistant estimators, new estimators result that are both highly resistant to outliers and highly efficient. M-estimation using the residuals from an initial highly resistant LTS fit, for example, leads to the S1S GM-estimator, which is highly resistant to both residual outliers and high leverage observations, and yet maintains an efficiency of about 95% relative to the OLS estimator. Computing M-estimates from the residuals from an LMS or S-estimation leads to the similarly efficient and robust MM-estimators.

EXAMPLE 4.1: Simulated Data

We now return to the simulated data first introduced in Chapter 3 that include various types of outliers. We explore six different regression estimates for each of the three "contaminated" data sets: an OLS estimate, an LAV estimate, an M-estimate (using Huber weights), a GM-estimate (specifically a Coakley-Hettmansperger estimator), an *S*-estimate, and an *MM*-estimate. The fitted lines for the estimates for each of the data sets are shown in Figure 4.1, with the "contaminating" observation identified.

Starting with the vertical outlier scenario, the substantive conclusions, at least in terms of the slope coefficients, are nearly identical regardless of which regression method is employed. In fact, aside from the OLS line, which stands apart only slightly in terms of its intercept, it is impossible to distinguish the lines for the other methods from each other. Moreover, although the OLS intercept is slightly smaller than the intercept for the others—indicating that the line is pulled toward the outlier—it is not so dissimilar from the others that it is problematic. The estimates for the "good" leverage point (B) are even more similar to each other. Aside from the LAV line, which falls just slightly above the others, all of the other regression lines are directly on top of each other. For the case of the "bad" leverage point (C), the various estimates differ to a greater degree, although the most marked differences are with respect to contrast between the OLS estimate and the others. As we saw in Chapter 3, the OLS regression line has been pulled toward the influential observation. None of the robust regression estimates is substantially influenced by the outlier, however. It is clear, then, that a more robust method should be favored in this last scenario. But what about the other two scenarios, for which very little difference was found between the estimates?

To answer this question, we turn to the distribution of the residuals to see whether the precision of the OLS estimates might be hindered. Recall that standard errors for the OLS estimates are smallest when there is less spread in the residuals. Figure 4.2, which shows the distribution of the residuals, suggests that the vertical outlier causes problems for the standard errors of the OLS estimators (see plot A). On the other hand, the residuals are very well behaved despite the addition of the "good" leverage point (plot B). The information in Figures 4.1 and 4.2 taken together suggest that OLS is only suitable for the data with the good leverage point. Given that all of the robust estimators tell the same general story, an efficient estimator like the *MM*-estimator would be a good choice for the other scenarios.

EXAMPLE 4.2: Multiple Regression Predicting Public Opinion

We now return to the cross-national public opinion data, continuing to focus on only the new democracies. Earlier, we used OLS to fit a model predicting public opinion from per capita GDP and the Gini coefficient. Diagnostics and preliminary analyses suggested that the model performed better if the Czech Republic and Slovakia were omitted (see Table 3.2). Recall that the OLS, including all observations, showed a statistically significant

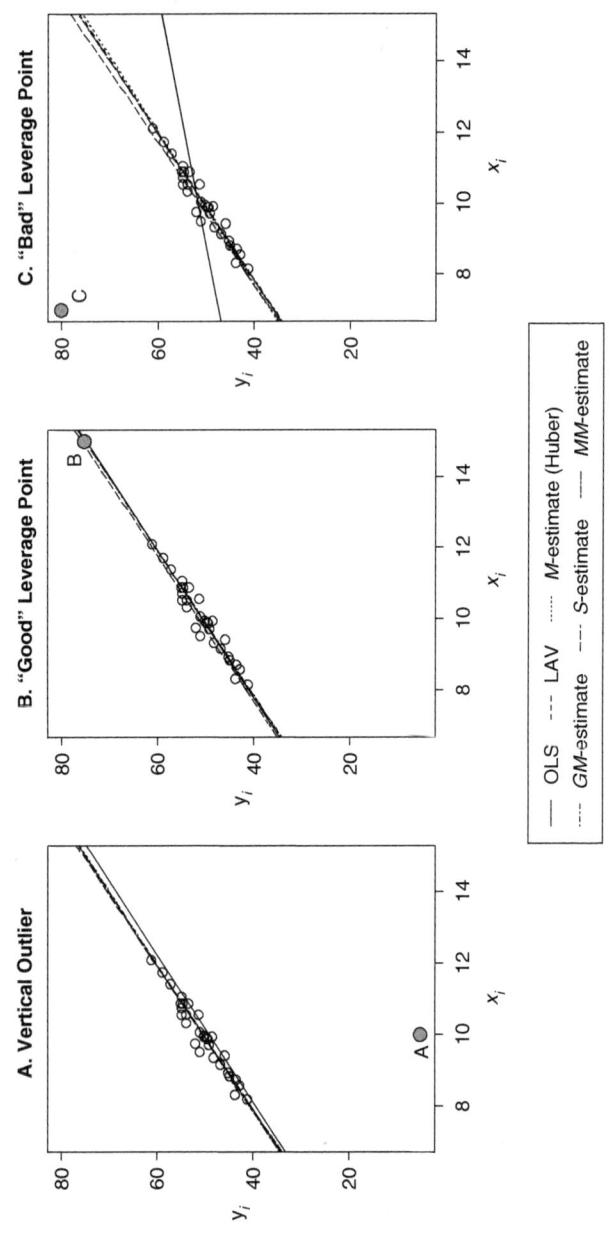

Figure 4.1   Various Regression Estimates for the Contrived Data Including Three Different Types of Outliers

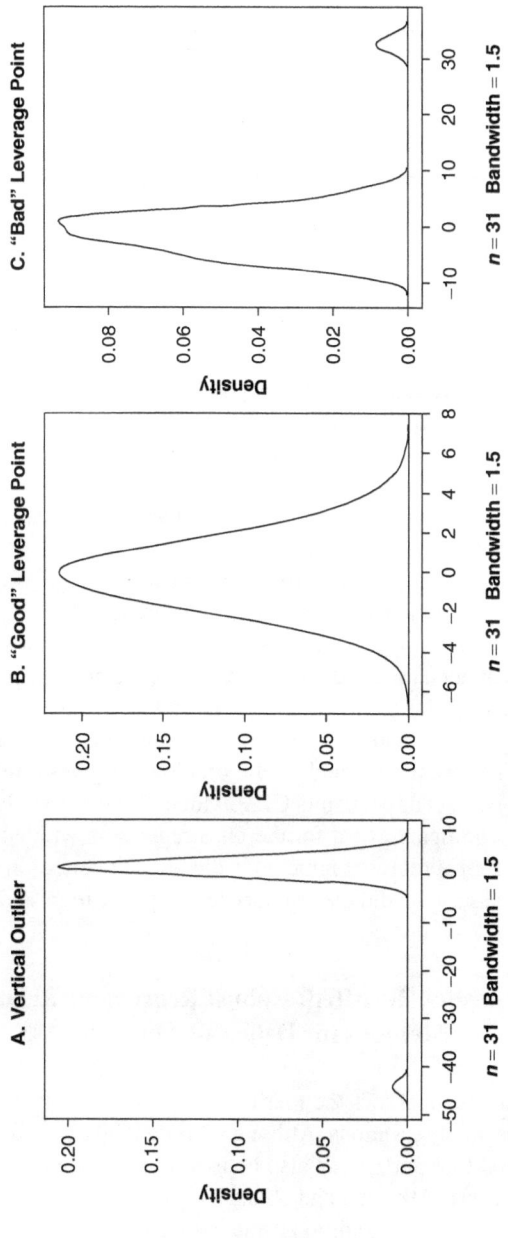

Figure 4.2   Density Estimates of the Residuals for the OLS Regression Fitted to the Three "Contaminated" Data Sets

TABLE 4.2
Robust Regression Models Fitted to the Public
Opinion Data, New Democracies

	LAV Regression	M-Estimation (Huber)	M-Estimation (Biweight)	MM-Estimation	Generalized M-Estimation (Coakley-Hettmansperger)
Intercept	−0.079	−0.063	−0.091	−0.097	0.939
Gini	0.0045	0.0039	0.0049	0.0051	0.0041
Per Capita GDP/1000	0.0059	0.0089	0.0052	0.0057	0.0065
n	26	26	26	26	26

positive effect of per capita GDP ($\hat{\beta} = 0.0175$) and a statistically insignificant effect for the Gini coefficient ($\hat{\beta} = 0.00074$). After the two outliers were removed, the coefficient for per capita GDP fell to about one third the size and was no longer statistically significant ($\hat{\beta} = 0.0063$), whereas the slope for the Gini coefficient became more than seven times as large and statistically significant ($\hat{\beta} = .00527$).

Table 4.2 gives the estimates from several robust regressions fitted to the same data. Although there are small differences between them, the M-estimators, MM-estimator, and GM-estimator tell a similar story regarding the effects of per capita GDP and the Gini coefficient. All of these methods give results similar to the OLS regression that omits the two outliers. The LAV regression also does a good job uncovering the relationship between the Gini coefficient and public opinion but gives a much smaller estimate for the effect of per capita GDP, which, in any event, does not have a statistically significant effect for the OLS regression without outliers. In summary, the robust regression methods did a much better job of handling the influential cases than did the ordinary least squares regression.

Diagnostics Revisited: Robust Regression-Related Methods for Detecting Outliers

The discussion above shows the merit of robust regression in limiting the impact of unusual observations. Although it is certainly sensible to see it as the final method to report, it can also be used in a preliminary manner as a diagnostic tool (see Atkinson and Riani 2000). In this respect, it can be a good complement to the traditional methods for detecting unusual cases that were discussed in Chapter 3.

A criticism of common measures of influence, such as Cook's D, is that they are not robust. Their calculation is based on the sample mean and covariance matrix, meaning that they can often miss outliers (see Rousseeuw and van Zomeren 1990). More specifically, Cook's D is prone to a "masking effect," where a group of influential points can mask the impact of each other. We have already seen that partial regression plots can be helpful for overcoming the masking effect with respect to individual coefficients. Information from the weights and residuals from robust regression can help combat the masking problem when assessing overall influence on the regression.

Index Plots of the Weights From the Final IWLS Fit

A straightforward way to use robust regression as a diagnostic tool involves the weights from the final IWLS fit. It is important to remember, however, that the meaning of the weights differs according to the model. Different methods could give quite different weights to observations depending on the type of unusualness. For M-estimation, the only thing that we can say about the weights is that they indicate the size of the residuals in an OLS fit, and thus whether or not the observation is a vertical outlier. Examined alone, they tell us nothing about leverage, and thus influence, because M-estimates do not take these into account when assigning the weights. GM-estimation, on the other hand, down-weights observations according to both the size of their residual in an OLS fit and their leverage, although an examination of the weights will not allow us to distinguish between the two aspects. Weights from the MM-estimate can provide a good indication of overall influence because of their highly resistant initial step.

To evaluate how the weights from various robust regressions perform, it is instructive to compare them to the Cook's distances and other outlier detection measures from an OLS regression fitted to the same data. Table 4.3 contains this information. We found in Chapter 3 that Cook's distances identified the Czech Republic and Slovakia as influential observations. All of the robust regression also detected the two discrepant observations, giving them comparatively low weight. In other words, the weights indicate the level of unusualness—the smaller the weight, the more unusual is the observation.

The evidence from Table 4.3 motivates the idea of plotting robust regression weights in an index plot in the same manner as is typically done with Cook's D. This has been done in Figure 4.3. Although all three robust regression methods identified the two most problematic observations, the GM-estimation gave nine other observations a weight less than .5, whereas neither of the other methods gave any other observation a weight less

TABLE 4.3
Diagnostic Information From OLS and Robust Regressions

Country	OLS Diagnostic Statistics			Final Weights From Robust Regressions			
	Cook's D	Hat Value	Studentized Residual	M-Estimate (Huber)	M-Estimate (Bisquare)	MM-Estimate	GM-Estimate
Armenia	0.0030	0.10	-0.27	1	0.81	0.87	0.35
Azerbaijan	0.0011	0.10	-0.17	1	0.97	0.98	0.86
Bangladesh	0.0012	0.18	-1.13	1	1	1	1
Belarus	0.0096	0.07	-1.63	1	1	1	1
Brazil	0.012	0.26	0.36	1	1	1	1
Bulgaria	0.0011	0.07	0.21	1	0.82	0.88	0.38
Chile	0.1350	0.24	1.14	0.72	0.55	0.72	0.18
China	0.0005	0.073	0.13	1	1	1	1
Croatia	0.0073	0.063	-0.56	1	1	1	1
Czech Republic	0.3629	0.17	2.60	0.22	0	0	0.62
Estonia	0.0155	0.04	-1.00	1	0.85	0.89	0.40
Georgia	0.0011	0.07	-0.21	1	0.98	0.99	1
Hungary	0.0273	0.098	-0.87	1	1	1	1
Latvia	0.0093	0.07	-0.59	1	1	1	1
Lithuania	0.0040	0.04	-0.51	1	0.99	1	1
Mexico	0.0003	0.17	0.06	1	1	1	1

	OLS Diagnostic Statistics			Final Weights From Robust Regressions			
Country	Cook's D	Hat Value	Studentized Residual	M-Estimate (Huber)	M-Estimate (Bisquare)	MM-Estimate	GM-Estimate
Moldova	0.0038	0.112	0.295	1	0.96	0.97	0.84
Nigeria	0.0504	0.14	0.95	1	0.89	0.93	0.47
Romania	0.0004	0.08	−0.11	1	0.91	0.94	0.55
Russia	0.0189	0.09	−0.76	0.87	0.68	0.77	0.27
Slovakia	0.6990	0.17	4.32	0.17	0	0	0.05
Slovenia	0.2691	0.24	−1.66	1	0.99	0.99	1
Taiwan	0.1296	0.15	−1.50	1	0.96	0.97	0.80
Turkey	0.0011	0.05	0.26	1	0.89	0.93	0.46
Ukraine	0.0024	0.1	−0.26	1	0.83	0.88	0.38
Uruguay	0.0059	0.07	0.46	0.95	0.65	0.78	0.24

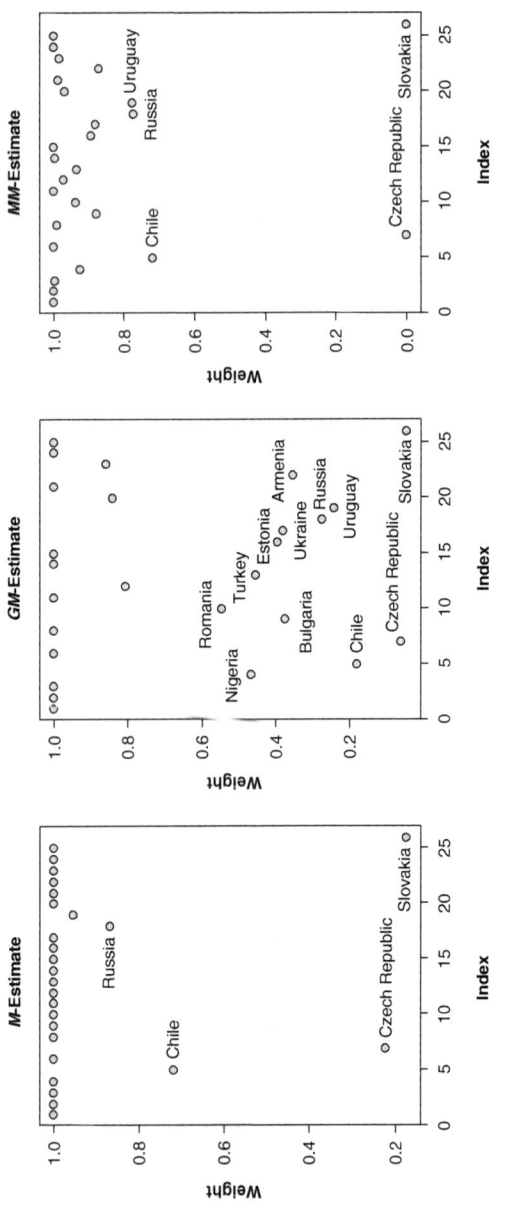

Figure 4.3   Index Plots of Final Weight From the IWLS Fit for Various Robust Regression Estimates

than .7. As said above, the uniqueness of the *GM*-estimate results because it considers the size of the residual *and* leverage.

RR-Plots ("Residual-Residual" Plots)

According to Rousseeuw and van Zomeren (1990:637), robust regression residuals are much better than OLS residuals for diagnosing outliers because the OLS regression "tries to produce normal-looking residuals even when the data themselves are not normal." With this in mind, Tukey (1991) proposed the RR-plot ("residual-residual" plot), which calls for a scatterplot matrix that includes plots of the residuals from an OLS fit against the residuals from several different robust regressions. If the OLS assumptions hold perfectly, there will be a perfect positive relationship, with a slope equal to 1 (called the "identity line"), between the OLS residuals and the residuals from any robust regression. Let the *i*th residual from the *j*th regression fit $\hat{\beta}_j$ be $e_{ij} = y_i - \mathbf{x}_i^T \hat{\beta}_j$, then

$$\left\| e_i(\hat{\beta}_1) - e_i(\hat{\beta}_2) \right\| = \left\| \hat{y}_i(\hat{\beta}_1) - \hat{y}_i(\hat{\beta}_2) \right\| = \left\| \mathbf{x}_i^T(\hat{\beta}_1 - \hat{\beta}_2) \right\|$$
$$\leq \|\mathbf{x}_i\| \left(\left\| \hat{\beta}_1 - \beta \right\| + \left\| \hat{\beta}_2 - \beta \right\| \right). \quad [4.25]$$

This implies that as *n* approaches ∞, the scatter around the identity line will get tighter and tighter if the regression assumptions are met. If there are outliers, the slope will be a value other than 1 because the OLS regression does not resist them whereas the robust regression does.

RR-plots for the public opinion data are shown in Figure 4.4. The broken line is the identity line; the solid line shows the regression of the residuals from the method on the vertical axis on the residuals from the method on the horizontal axis. The plots in the first column are of most interest because they show the regression of the OLS residuals on the residuals from various robust regression methods. The fact that the two lines are far apart from each other in all of these plots indicates that the OLS estimates were highly influenced by the outliers; the Czech Republic and Slovakia have much smaller residuals for the OLS regression, indicating that they are quite influential. Turning to the other plots, we notice that the residuals from the various robust regressions are very similar to each other, especially with respect to the *MM*-estimates and *GM*-estimates, which are nearly identical.

Robust Distances

We can also consider diagnostic methods that pertain only to a robust regression. For example, Rousseeuw and van Zomeren (1990) claim that a plot of the robust residuals against robust distances, the latter being based on the Mahalanobis distance but defined by a robust covariance matrix, is

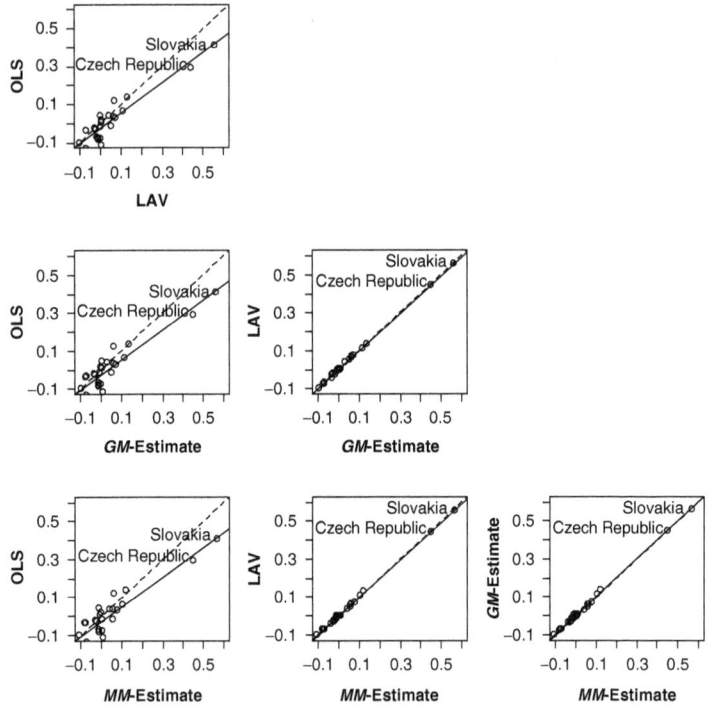

Figure 4.4   RR-Plots for the Regression of Public Opinion Regressed on Per Capita GDP and the Gini Coefficient, New Democracies

able to detect multiple outliers better than traditional methods (see also Cook and Hawkins 1990; Ruppert and Simpson 1990; and Kempthorne and Mendel 1990 for debate about this topic). Efficiency is not a concern for these diagnostics, so the residuals from the highly resistant LMS or LTS regressions are often employed.

The Mahalanobis distance measures how far an observation x_i is from the center of the cloud of points defined by the data set **X.** It is defined by

$$\text{MD}_i = \sqrt{\left(\mathbf{x}_i - \bar{\mathbf{x}}\right)\text{cov}(\mathbf{X})^{-1}(\mathbf{x}_i - \bar{\mathbf{x}})^T}, \qquad [4.26]$$

where $\bar{\mathbf{x}}$ is the centroid of **X** and $\text{cov}(\mathbf{X})$ is the sample covariance matrix. Outliers can influence the mean and covariance matrix, and thus they will not necessarily be detected by the MD_i. As a result, Rousseeuw and van Zomeren's (1990) robust distances RD_i are defined by replacing $\text{cov}(\mathbf{X})$ and $\bar{\mathbf{x}}$ with

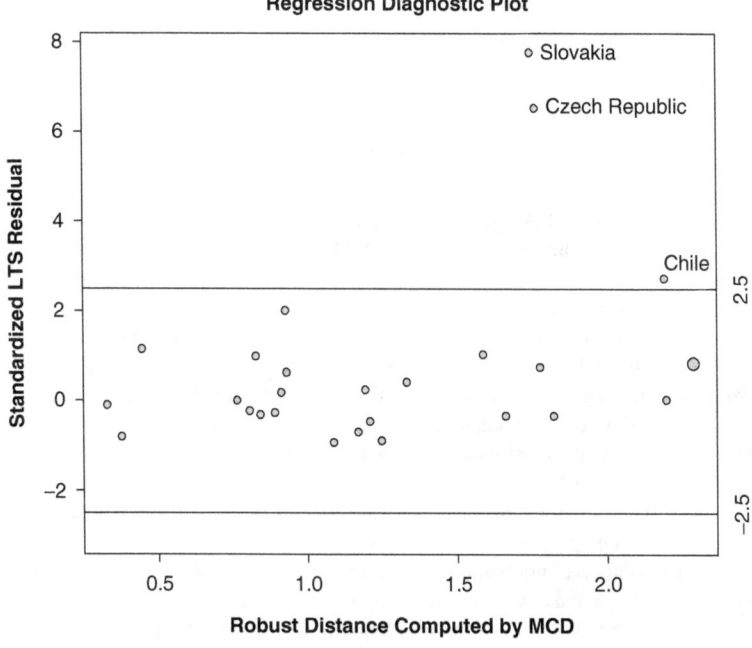

Figure 4.5   Plot of Robust Residuals (From LTS Fit) Against Robust Distances

the more robust center and covariance matrix from the minimum volume ellipsoid estimator (see Rousseeuw 1985 for more details). Usual practice is to identify standardized robust residuals as problematic if they are $e' \geq |2.5|$. Similarly, robust distances are identified as having high leverage if $RD_i > 0.975$ percent point of the chi-squared distribution with degrees of freedom equal to the number of parameters estimated in the model.

Rousseeuw and van Zomeren's regression diagnostic plot for the public opinion data is shown in Figure 4.5. Plotted against the robust distances are the standardized residuals from a LTS regression. Although the robust distances indicate that no cases have unusually high leverage, the robust residuals suggest that three cases are outliers. Following in line with the rest of the analyses that we have done so far, Slovakia and the Czech Republic are two of these. The third observation, which is just barely past the rule of thumb cutoff, is Chile.

As well as the methods discussed above, the traditional diagnostic plots for identifying outliers (discussed in Chapter 3) can be extended to robust regression models. Given that they are generally interpreted in the same way as for the OLS fit, they're not discussed here. For more information on these diagnostics, see McKean and Sheather (2000). Other techniques related to robust regression can also be seen in Fung (1999) and Pena and Yohai (1999).

Notes

1. Other names for LAV regression are least absolute deviations (LAD) regression and minimum sum of absolute errors (MSAE) regression (Birkes and Dodge 1993).
2. Related methods not discussed in this book because of their limited use are the least-trimmed median estimators and the least-trimmed difference estimators. Both of these have breakdown points of $BDP = 0.5$, but their relative efficiency is less than 67%. For more information, see Croux et al. (1994) and Stromberg et al. (2000).
3. These estimators are sometimes referred to as *trimmed-mean* estimators. They can also be modified so that they have a bounded influence function (see De Jongh, De Wet, and Welsh 1988).
4. The LMS estimator should not be confused with Siegel's (1982) repeated median (*RM*). Although proposed as a robust estimator quite early, the *RM* estimator has the severe limitation of not being affine-regression equivariant (i.e., coefficient estimates do not behave as expected when the predictors are rescaled or combined in linear ways) for high-dimensional problems. It will not be discussed any further because of this limitation.
5. IRLS is also often referred to as *iterative weighted least squares* (IWLS).
6. This assumes tuning constants of $c = 1.345$ for Huber weights and $c = 4.685$ for biweights. See the earlier discussion of *M*-estimation of location for more details.
7. Other methods, such as LMS estimation (Rousseeuw 1984) and RM estimation (Siegel 1982), have also been proposed for the initial stage.

5. STANDARD ERRORS FOR ROBUST REGRESSION

Analytical standard errors are easily calculated for some, but not all, types of robust regression.[1] Nonetheless, even when analytical standard errors can be calculated, they are not reliable for small samples. As a result, it is often desirable to use bootstrapping to calculate standard errors. As a result, this chapter starts with a brief discussion of asymptotic standard errors, and then continues by exploring various types of bootstrapped standard errors and confidence intervals.

Asymptotic Standard Errors
for Robust Regression Estimators

Analytical standard errors are obtainable for estimators of S and M class (including generalized versions and MM-estimators). For all of these estimators, asymptotic standard errors (ASE) are given by the square root of the diagonal entries of the estimated asymptotic covariance matrix for the coefficient estimates from the final IRLS fit, $V(\hat{\beta}) = s_e^2 (\mathbf{X}^T \mathbf{W} \mathbf{X})^{-1}$ (Draper and Smith 1998:575; see also Hill and Holland 1977 and Birch and Agard 1993),[2] where $\mathbf{W}$ is the final weight matrix and is S_e^2 the variance of the residuals. The variance of the residuals is defined by

$$s_e^2 = \frac{1}{n-p} \mathbf{y} \left[\mathbf{W}^{-1} - \mathbf{W}^{-1} \mathbf{X} \left(\mathbf{X}^T \mathbf{W}^{-1} \mathbf{X} \right)^{-1} \mathbf{X}^T \mathbf{W}^{-1} \right] \mathbf{y}, \qquad [5.1]$$

where p is the number of parameters in the model. These estimates are routinely provided in the regression output for most statistical packages.

The ASEs can be considered reliable if the sample size n is sufficiently large relative to the number of parameters to estimate (Yohai 1987). If n is small (e.g., less than 40), however, they cannot be trusted (Li 1985; Huber 2004:164). Other evidence also suggests that their reliability decreases as the proportion of influential cases increases (Stromberg 1993). It is recommended, then, that ASEs be used only with large sample sizes. When n is small, bootstrapping provides an alternative way to obtain standard errors or confidence intervals.[3]

Bootstrapped Standard Errors

First proposed by Efron (1979; see also Efron 1981), bootstrapping can be used to compute standard errors and confidence intervals for a statistic that does not have an easily derivable asymptotic standard error, or if the assumptions necessary for its use have been violated. Bootstrap standard errors are found by repeatedly sampling from the original sample. Although bootstrapping for regression is not desirable if the OLS assumptions are met—in such cases, OLS estimates are the most efficient of unbiased estimates[4]—if the assumptions are not met, bootstrapping can be useful. With respect to robust regression, bootstrapping is most useful when n is small because the ASEs cannot be trusted in these conditions. Bootstrapping for robust regression can be done in two ways: *random-x bootstrapping* or *fixed-x bootstrapping*.

Random-*x* Bootstrapping

Random-*x* bootstrapping resamples individual *observations* from the data set. In other words, it samples rows from the data matrix. This is an appropriate way to proceed when the *regressors are random*—that is, a different result could be expected with each new random sample—as in the case of large-scale survey data (Mooney and Duval 1993:17). Assume a data set with one independent variable x_i and a dependent variable y_i. The steps to obtain random-*x* bootstrapped standard errors are rather simple:

1. From the original sample data (x, y), randomly select $B = 1, \ldots, R$ samples of size *m* with replacement (x^*, y^*). Call these *R* samples the bootstrap samples, *B*. With small samples, it is common to select $R = n^n$ bootstrap samples. Selecting n^n samples is impractical for large samples (e.g., it would be foolish to calculate $1,000^{1,000}$ bootstrap samples), however, and thus 1,000 bootstraps is typically considered acceptable.
2. Calculate the robust regression estimates $\hat{\beta}_0^*$ and $\hat{\beta}_1^*$ for each of the *R* bootstrap samples.
3. Use the empirical distributions of $\hat{\beta}_0^*$ and $\hat{\beta}_1^*$ to calculate standard errors for $\hat{\beta}_0$ and $\hat{\beta}_1$, applying principles of inference similar to the classical methods for random samples from a larger population. In other words, the bootstrapped standard errors are calculated from the distribution of the statistic from the bootstrap samples rather than from the statistic's unknown sampling distribution.

Fixed-*x* Bootstrapping

Fixed-*x* bootstrapping is an appropriate alternative to random-*x* bootstrapping when the explanatory variables (i.e., the model matrix **X**) are assumed to be fixed. This method is a bit more complicated than random-*x* bootstrapping because it *resamples the residuals* from the regression model rather than the observations themselves. It proceeds as follows:

1. Treat the fitted values $\hat{y}_i$ from the original robust regression as the expectation of the response from the bootstrap.
2. In the usual manner, calculate the residuals, $e_i = y_i - \hat{y}_i$, from the regression model.
3. Randomly select $B = 1, \ldots, R$, samples of size *n* with replacement *from the residuals* e_i. As in the case of the random-*x* bootstrap, $R = 1,000$ is typical. Call the resampled residuals $\hat{e}_{Bi}^*$.
4. Add the resampled residuals $\hat{e}_{Bi}^*$ to the fitted values from the regression, producing the fixed-*x* bootstrap sample, $y_{\hat{\beta}i}^* = \hat{y}_i + \hat{e}_{Bi}^*$, creating *R* sets of bootstrap fitted values.

5. Regress each set of the R bootstrap fitted values $y^*_{\hat{\beta}}$ on the fixed model matrix $\mathbf{X}$, producing R sets of regression coefficients.
6. Create confidence intervals (or standard errors) from the empirical distribution of the bootstrap replicates of the coefficients.

Constructing Confidence Intervals

Regardless of whether random-x or fixed-x bootstrapping is employed, the mean of the bootstrapped regression coefficients is

$$\bar{\hat{\beta}}^* = \hat{E}^*\left(\hat{\beta}^*\right) = \frac{\sum\limits_{B=1}^{R} \hat{\beta}^*_B}{R}. \qquad [5.2]$$

The estimated bootstrap variance of $\hat{\beta}^*$ is

$$\hat{V}^*\left(\hat{\beta}^*\right) = \frac{\sum\limits_{B=1}^{R} \left(\hat{\beta}^*_B - \bar{\hat{\beta}}^*\right)^2}{R-1}. \qquad [5.3]$$

Three types of confidence intervals are commonly considered: (1) normal-theory intervals, (2) percentile intervals, and (3) bias-corrected percentile intervals. Graphical examination of the bootstrap distribution aids in determining which confidence interval to employ.

If the mean of the bootstrap sampling distribution is unbiased, that is, $\bar{\hat{\beta}}^* = \hat{\beta}$, normal-theory and percentile intervals can be considered. Following from the notion that many statistics are asymptotically normally distributed, *normal-theory intervals* can be justified when the bootstrap sampling distribution is approximately normal. A $100(1 - \alpha)\%$ confidence interval is calculated in the standard way:

$$\beta = \hat{\beta} \pm z_{\alpha/2} S\hat{E}^*\left(\hat{\beta}^*\right), \qquad [5.4]$$

where the bootstrap estimate of the standard error, $S\hat{E}^*\left(\hat{\beta}^*\right)$, is the standard deviation of the bootstrap sampling distribution.

If the mean of the bootstrap statistic is unbiased but its bootstrap sampling distribution is not normally distributed, *percentile intervals* are more appropriate. These are calculated by first ordering the bootstrap statistics from smallest to largest value: $\hat{\beta}^*_{(1)} \le \hat{\beta}^*_{(2)} \le \cdots \le \hat{\beta}^*_{(R)}$. The confidence interval is then bounded at the $\alpha/2$ and $1 - \alpha/2$ quantiles of the bootstrap sampling distribution $\hat{\beta}^*_{(\alpha/2)} < \beta < \hat{\beta}^*_{(1-\alpha/2)}$, where $100(1 - \alpha)\%$ is the desired level of confidence.

Normal-theory and percentile confidence intervals work well when the bootstrap estimate is unbiased, which is often the case with large samples. If the estimate is substantially biased—which is common with small samples—adjustments must be made. To this end, *bias corrected (BC) confidence intervals* employ a normalizing transformation using two correction factors, Z and A. Z is defined as

$$Z = \Phi^{-1} \left[\frac{\#_{b=1}^{R} \left(T_b^* \leq T \right)}{R+1} \right], \qquad [5.5]$$

where Φ is the standard normal density function, and $\#_{b=1}^{R} \left(T_b^* \leq T \right) / (R+1)$ is the adjusted proportion of the bootstrap replicates below the original sample estimate $\hat{\beta}$. The A correction factor is defined by

$$A = \frac{\sum_{i=1}^{n} \left(T_{(-i)} - \bar{T} \right)^3}{6 \left[\sum_{i=1}^{n} \left(T_{(-i)} - \bar{T} \right)^2 \right]^{3/2}}, \qquad [5.6]$$

where $T_{(-i)}$ represents the value of T from jackknife resampling[5] that removes the ith observation, and is the average of the n jackknife values, that is, $\bar{T} = \sum_{i=1}^{n} T_{(-i)}/n$. The lower and upper bounds of the *BC confidence interval* are then calculated as follows:

$$BC_{lower} = \Phi \left[Z + \frac{Z - z_{\alpha/2}}{1 - A \left(Z - z_{\alpha/2} \right)} \right]$$

$$BC_{upper} = \Phi \left[Z + \frac{Z + z_{\alpha/2}}{1 - A \left(Z + z_{\alpha/2} \right)} \right] \qquad [5.7]$$

For more details of the BC confidence intervals, see Efron and Tibshirani (1993: section 14) and Davidson and Hinkley (1997:103–107).

EXAMPLE 5.1: The Impact of Democracy on the Effects of Income Inequality and Per Capita GDP on Public Opinion

Continuing with the cross-national public opinion data, we now explore a model that predicts public opinion from the Gini coefficient, per capita GDP, and democracy. Preliminary analysis indicated an interaction between the Gini coefficient and democracy, so it is included in the model. Diagnostics not shown found that the Czech Republic and Slovakia continue to have unusual influence on the regression estimates, suggesting that robust regression is a good alternative to OLS. Coefficients and standard errors for various models are reported in Table 5.1. The estimates included in the table are from an OLS regression with all observations included, an OLS regression with the two outliers removed, an M-estimation, and an MM-estimation.

Although asymptotic standard errors are reported for the robust regression estimates, they are not trustworthy because of the small sample size ($n = 48$), so bootstrap standard errors were also calculated. The observations represent countries—and thus the model matrix can reasonably be considered as fixed—so fixed-x resampling is employed to find bootstrap confidence intervals. A total of $R = 1,000$ bootstrap samples of the residuals were taken from the original robust regression. Figure 5.1 presents diagnostics for the bootstrapping—a histogram and quantile comparison plot of the bootstrap distribution—only for the M-estimate for the democracy coefficient. The broken vertical line in the middle of the histogram represents the mean of the bootstrap estimates, indicating that it is nearly identical to the coefficient from the robust regression ($\hat{\beta} = 0.374$). In other words, the bootstrapped estimate is unbiased. The plots indicate that the distribution is slightly heavy-tailed but is otherwise roughly normal, suggesting the use of percentile confidence intervals. Diagnostics for the other coefficients led to similar conclusions, and thus all of the bootstrapped standard errors in Table 5.1 are based on percentile confidence intervals.

Most obvious in Table 5.1 is the continuing detrimental effect of the two outliers on the OLS estimates. None of the OLS estimates is statistically significant when the outliers are present. When they are removed, except for per capita GDP, which remains statistically insignificant ($p = .82$), the magnitudes of all the coefficients increase dramatically and become statistically significant (the largest $p = .0032$). Differences in the two robust regression methods are also noticeable. Contrary to the previous examples, we see that M-estimates are more vulnerable than MM-estimates to unusual observations. Because the M-estimates down-weight observations only according to the size of their residuals, they fail to consider leverage. The Czech Republic and Slovakia have large hat values along with large residuals. This high leverage has caused trouble for the M-estimates, especially with respect to the Gini coefficient, the effect of which is much smaller than it is for the OLS without the outliers or for the MM-estimation. The bounded influence MM-estimates are nearly identical to the OLS estimates with the two outliers removed, reflecting that the two observations were both given a weight of 0. The third noteworthy observation is the similarity in the asymptotic standard errors and the bootstrap standard errors. For none of the coefficients do the differences in the standard errors substantially affect tests of significance, suggesting that the asymptotic standard errors are fine in this case.

TABLE 5.1
Estimates and Standard Errors for OLS and Robust Regression Predicting Public Opinion, All Countries

	OLS (All Cases)		OLS (Outliers Removed)		M-Estimation (Huber)			MM-Estimation		
	$\hat{\beta}$	SE$(\hat{\beta})$	$\hat{\beta}$	SE$(\hat{\beta})$	$\hat{\beta}$	ASE$(\hat{\beta})$	Boot SE (fixed-x) R = 1,000	$\hat{\beta}$	ASE$(\hat{\beta})$	Boot SE (fixed-x) R = 1,000
Intercept	1.162	0.098	0.939	0.063	1.019	0.064	0.061	0.952	0.061	0.061
Democracy	0.114	0.182	0.467	0.115	0.374	0.119	0.112	0.475	0.113	0.119
Gini Coefficient	−0.0006	0.002	0.0049	0.0016	0.0029	0.0016	0.0015	0.0046	0.0015	0.0015
Per Capita GDP/1000	0.0039	0.004	0.0005	0.002	0.0010	0.0025	0.0023	0.0002	0.0023	0.0024
Democracy × Gini	−0.0028	0.005	−0.010	0.002	−0.008	0.003	0.0030	−0.011	0.0029	0.0030
n	48	46	48							

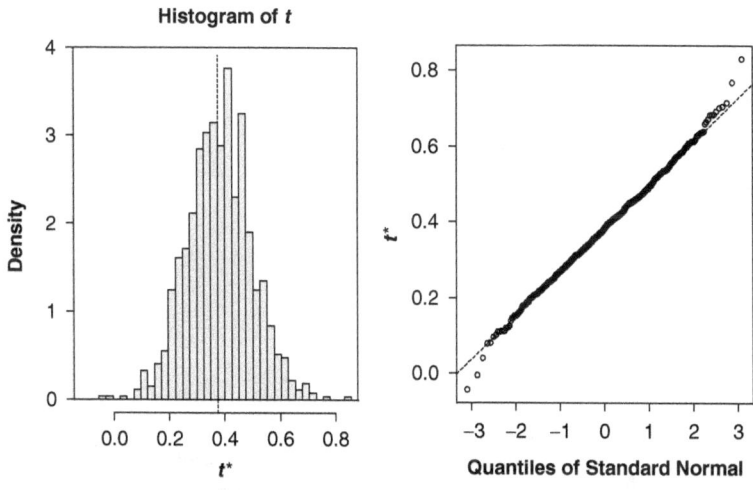

Figure 5.1   Diagnostic Plots of Bootstrap Replicates for the Democracy
Coefficient From *M*-Estimation Robust Regression

To more clearly see how different the estimates from these models are, Figure 5.2 displays fitted values from the OLS regression and the *MM*-regression showing the interaction between democracy and the Gini coefficient. To calculate these fitted values, democracy and the Gini coefficient were allowed to vary through their ranges in the regression equation, and per capita GDP was set to its mean. In other words, the fitted regression lines show the combined effects of Gini and democracy for a country with a typical per capita GDP. Considering that none of the effects is statistically significant for the OLS regression, it isn't surprising that the effect display also shows little pattern. In the case of *MM*-estimation, on the other hand, the interaction is clearly very strong. For old democracies, public opinion is less favorable of pay equality as income inequality (the Gini coefficient) rises. The opposite is true for new democracies. Although in different directions, the Gini coefficient has a strong impact for both new and old democracies.

Notes

1. As said earlier, standard errors for robust regression are something entirely different from the more commonly used "robust standard errors" used to compensate

78

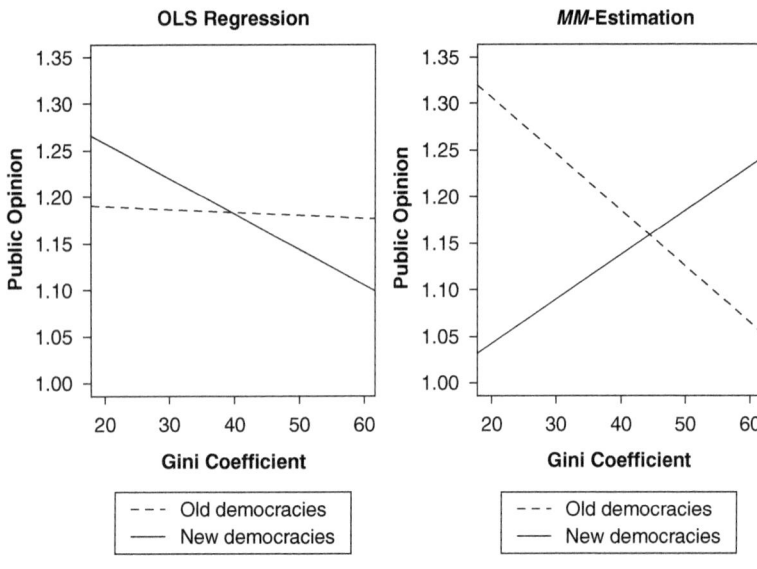

Figure 5.2   Effect Displays for the Interaction Between Democracy and the Gini Coefficient on Public Opinion Toward Pay Inequality, OLS Regression Model (Including Outliers), and *MM*-Estimation

for an unknown pattern of heteroscedasticity. Known by many different names—such as White, Eicker, and Huber standard errors—robust standard errors, and more generally "sandwich estimators," are calculated without making any changes to the OLS regression itself. For more general information on robust standard errors, see White (1980). For a good discussion of the performance of several types of robust standard errors, see Long and Ervin (2000).

2. For a brief, but very good, description of likelihood methods for calculating the standard errors, see Western (1995).

3. If the sample size is too small, bootstrapping can also fail. Chernick (1999:151) recommends a minimum sample size of 30. Bootstrapping can also fail for dependent data or if a significant proportion of the data is missing not at random (see Davidson and Hinkley 1997:37–54; Chernick 1999:102–105).

4. With large sample sizes, bootstrapping a linear model results in standard errors that approximate the usual standard errors.

5. Jackknife resampling differs from bootstrapping in that rather than take random samples of the data with replacement, it resamples the data (typically *n* times) by randomly removing a single observation (for more details, see Davidson and Hinkley 1997:113–118).

6. INFLUENTIAL CASES IN GENERALIZED LINEAR MODELS

The generalized linear model (GLM) extends from the general linear model to accommodate dependent variables that are not normally distributed, including those that are not continuous. This chapter starts with a brief description of the GLM. It then provides a brief discussion of diagnostic methods for detecting unusual cases in the GLM. It ends with an introduction to robust generalized linear models, providing empirical examples for logistic and Poisson regression models.

The Generalized Linear Model

I provide only a basic description of the GLM, emphasizing information that is necessary to understand robust generalized linear models. For a more extensive and detailed description of GLMs, see McCullagh and Nelder's (1989) classic book on the topic (see also Dobson 1990; Fahrmeir and Tutz 2001; and Lindsey 1997 for good general treatments of the GLM). For other discussions of the GLM geared toward social scientists, there are three books in the present series (Gill 2001; Dunteman and Ho 2005; Liao 1994).

Recall that the linear model is written as

$$y_i = \sum_{j=1}^{k} x_{ij}\beta_j + \varepsilon_i, \qquad [6.1]$$

where y is assumed to be linearly related to the xs, and the errors are assumed to be uncorrelated, have constant variance, and be normally distributed. In other words, the linear model represents the conditional mean of y given the xs as

$$\mu_i = \sum_{j=1}^{k} x_{ij}\beta_j. \qquad [6.2]$$

The generalized linear model loosens these assumptions to predict the conditional mean of a dependent variable with any exponential distribution, taking the following general form

$$f(y_i; \theta_i; \varphi) = \exp\left[\frac{y\theta - b(\theta)}{a(\varphi)} + c(y, \varphi)\right], \qquad [6.3]$$

where θ is the *canonical parameter* that represents the estimate of location, and φ is the *dispersion parameter* that represents the scale. In other words, the GLM allows the distribution of y to take the shape of many different exponential families:

$$y_i \mid xs \sim \begin{cases} \text{Gaussian} \\ \text{Binomial} \\ \text{Poisson} \\ \text{gamma} \\ \text{etc.} \end{cases}$$

The exponential family is defined by the a, b, and c functions in Equation 6.3. The assumption of linearity remains for the GLM but it is with respect to a *linear predictor* η rather than to y itself

$$\eta_i = \sum_{j=1}^{k} x_{ij}\beta_j. \qquad [6.4]$$

In other words, the canonical parameter θ in Equation 6.3 depends on the *linear predictor*. More specifically, the conditional mean μ_i of the dependent variable is linked to this linear predictor through a transformation, called the *link function* $g(.)$:

$$g(\mu_i) = \eta_i \qquad [6.5]$$

The link function must be monotonic and differentiable, and take any value (positive or negative) that ensures the linear dependence of η on the explanatory variables. An OLS regression is fitted when the identity link and the Gaussian family are specified. Any other link function results in a nonlinear relationship between the expectation of the dependent variable y_i and the independent variables x_{ij}. Table 6.1 displays some important families included in the GLM framework and some associated link functions.

Maximum likelihood estimates for GLMs are found by regarding Equation 6.3 as a function of the parameters β. Typically, this means maximizing the log-likelihood function with respect to β:

$$l(\beta) = \log L(\beta) = \log \prod_{i=1}^{n} f(y_i; \mu_i)$$

$$= \log \prod_{i=1}^{n} f(y_i; \mathbf{x}_i, \beta) = \sum_{i=1}^{n} \log f(y_i; \mathbf{x}_i, \beta) \qquad [6.6]$$

Maximum likelihood estimates can be obtained using the Newton-Raphson method or iteratively reweighted least squares (see Nelder and Wedderburn 1972; McCullagh and Nelder 1989). For IRLS estimation of GLMs, the

TABLE 6.1

Important Exponential Families and Their Link Functions

Distribution	Range of μ	Link Function, (g)	
Normal	$(-\infty, +\infty)$	Identity link	$g(\mu) = \mu$
Binomial	$(0, 1)$	Logit link	$g(\mu) = \log\left[\mu/(1 - \mu)\right]$
	$(0, 1)$	Probit link	$g(\mu) = \Phi^{-1}(\mu)$
Poisson	$(0, \infty)$	Log link	$g(\mu) = \log(\mu)$
Gamma	$(0, \infty)$	Reciprocal link	$g(\mu) = \mu^{-1}$
	$(0, \infty)$	Log link	$g(\mu) = \log(\mu)$

dependent variable is not y itself, but the *adjusted dependent variable z*, which is a linearized form of the link function applied to y. We start by defining the linear predictor for the first iteration

$$\underset{(n \times 1)}{\hat{\eta}^{(0)}} = \underset{(n \times p)}{X^T} \underset{(p \times 1)}{\beta^{(0)}}$$ [6.7]

with initial fitted values of $\hat{\mu}^{(0)}$ resulting from $g^{-1}(\hat{\eta}^{(0)})$. We then define z as

$$z^{(0)} = \hat{\eta}^{(0)} + \left(\left.\frac{\partial \eta}{\partial \mu}\right|_{\hat{\mu}^{(0)}}\right)\left(y - \hat{\mu}^{(0)}\right).$$ [6.8]

The *quadratic weight matrix* to be used in the IRLS is defined by

$$W_{(0)}^{-1} = \left(\left.\frac{\partial \eta}{\partial \mu}\right|_{\hat{\mu}^{(0)}}\right)^2 V(\mu)|_{\hat{\mu}^{(0)}},$$ [6.9]

where $V(\mu)$ is the variance function defined at $\hat{\mu}^{(0)}$. Both z and $W_{(0)}$ depend on the current fitted value, and thus an iterative process is needed to find a solution. We first regress $z^{(0)}$ on the xs with weight $W_{(0)}$ to find new estimates of the regression coefficients $\hat{\beta}^{(1)}$, and from these a new estimate of the linear predictor. Using the new estimates of z and W, the estimation process is continually repeated until convergence, resulting in normal equations of the general form

$$\hat{\beta} = \left(\mathbf{X}^T \mathbf{W} \mathbf{X}\right)^{-1} \mathbf{X}^T \mathbf{W} \mathbf{z},$$ [6.10]

where $\mathbf{z}$ represents the adjusted dependent variable transformed by the link function and $\mathbf{W}$ is the final weight matrix. GLMs are further extended by *quasi-likelihood estimation*, which, along with the usual specification of the link function, allows specification of the dispersion parameter φ instead of the entire distribution of y (see Wedderburn 1974 for more details).

The deviance of the model parallels the residual sum of squares for least squares regression in that it compares the model under investigation with the saturated model β_S for the data. A saturated model with n coefficients for the n observations matches the data exactly, meaning that it achieves the highest possible likelihood. The likelihood of this saturated model provides a baseline to which the likelihood of a less than saturated model can be compared. The deviance measures the discrepancy in fit between the two models. More specifically, it is twice the difference between the log likelihood of the saturated model and the log likelihood achieved by the model under investigation

$$D(\beta, \mathbf{y}) = 2[\log L(\beta_S)] - 2[\log L(\beta)]$$
$$= -2[\log L(\beta)]. \qquad [6.11]$$

The deviance plays an important role in assessing the fit of the model and in statistical tests for parameters in the model, and also provides one method for calculating residuals that can be used for detecting outliers.

Detecting Unusual Cases in Generalized Linear Models

As for OLS regression, unusual cases can distort estimates for GLMs. For some models, such as the binary logit and probit models, the impact of unusual cases is usually less severe because the dependent variable has only two possible values, but it is still possible for such observations to affect the regression estimates. For other models, like Poisson regression, highly unusual values of the dependent variable are more likely. It is important, then, to explore for outliers in GLMs. Many diagnostic tools for OLS regression have been adapted for the GLM, and those for assessing unusual observations are quite effective.

Residuals From the GLM

Residuals from GLMs can be defined in several ways. Some of these include the response residuals, which are simply the difference between the observed value of y and its fitted value, $y_i - \hat{\mu}_i$; the *deviance residuals*, which are derived from the case-wise components of the deviance of the model; and the *working residuals*, which are the residuals from the final iteration of weighted least squares. There are also approximations of the studentized residuals. This book is most concerned with *Pearson residuals* because they play a central role in many robust GLM models. Pearson residuals are simply the response residuals scaled by the standard deviation of the expected value:

$$e_{\text{Pearson}_i} = \frac{y_i - \hat{\mu}_i}{\sqrt{V(\hat{\mu})}}.$$ [6.12]

For more details of the relative merits of the various types of residuals, see Gill (2001). Each has its uses, and none of them is best for all purposes.

Hat Values and Leverage

As with OLS regression, leverage in the GLM is assessed by the *hat values* h_i, which are taken from the final IWLS fit. Unlike in linear regression, however, the hat values for GLMs depend on the values of y *and* the values of x. Following from Pregibon (1981), the hat matrix is defined by

$$\mathbf{H} = \mathbf{W}^{1/2}\mathbf{X}(\mathbf{X}^T\mathbf{W}\mathbf{X})^{-1}\mathbf{X}^T\mathbf{W}^{1/2},$$ [6.13]

where $\mathbf{W}$ is the weight matrix from the final iteration of the IWLS fit. This differs from the general form of $\mathbf{H}$ (Equation 3.7) by replacing $\mathbf{X}$ with $\mathbf{W}^{1/2}\mathbf{X}$. Doing so allows for a change in the variance of $\mathbf{y}$, and thus the hat values depend on both $\mathbf{y}$ and $\mathbf{X}$ (see McCullagh and Nelder 1989:405).

Assessing Influence

Following the linear model, DFBETAs and Cook's distances are helpful for detecting influence in GLMs. DFBETAs are calculated by finding the difference in an estimate before and after a particular observation is removed, $D_{ij} = \hat{\beta}_j - \hat{\beta}_{j(-i)}$, for $i = 1, \ldots, n$ and $j = 0, 1, \ldots, k$. An approximation of Cook's D measure of influence is also available:

$$D_i = \frac{e^2_{\text{Pearson}_i}}{\hat{\varphi}(k+1)} \times \frac{h_i}{1 - h_i},$$ [6.14]

where $\hat{\varphi}$ is the estimated dispersion of the model and k is the number of parameters being estimated excluding the constant (see Fox 2002).

Robust Generalized Linear Models

Methods for robust estimation of GLMs have developed much more slowly than robust methods for linear regression. Although there were several early attempts to make logistic regression more robust (e.g., Pregibon 1981; Copas 1988; Carroll and Pederson 1993; Bianco and Yohai 1996), the extension to other GLMs was seldom considered. Still today there are very few statistical programs that have routines for robust GLMs, and those that do are usually limited to the logit and Poisson model.

M-Estimation for GLMs

As with the linear model, the most widely used robust methods for the GLM are based in some way on *M*-estimation. Like early *M*-estimators for linear regression, many early attempts at *M*-estimation for GLMs suffered from an unbounded influence function (see Stefanski, Carroll, and Ruppert 1986; Kunsch, Stefanski, and Carroll 1989). Often, the resulting estimators were also undesirable because they were Fisher inconsistent.[1] In recent years, however, consistent bounded influence methods based on quasi-likelihood estimation have developed. One of these methods is due to Cantoni and Ronchetti (2001).[2]

Cantoni and Ronchetti's estimator evolved from the quasi-likelihood generalized estimating equations of Preisser and Qaqish (1999):

$$\sum_{i=1}^{n} \frac{\partial}{\partial \beta} Q(y_i; \mu_i) = \sum_{i=1}^{n} \frac{(y_i - \mu_i)}{V(\mu_i)} \mu_i' = 0, \qquad [6.15]$$

where $\mu_i' = \frac{\partial}{\partial \beta} \mu_i$ and $Q(y_i; \mu_i)$ is the quasi-likelihood function. The solution is an *M*-estimator defined by the score function

$$\Psi(y_i; \mu_i) = \frac{(y_i - \mu_i)}{v(\mu_i)} \mu_i'. \qquad [6.16]$$

Unfortunately, this estimator is limited for robust regression because its influence is proportional to Ψ, and thus unbounded.

Cantoni and Ronchetti follow the logic of Mallow's *GM*-estimates for regression (see Chapter 4) to improve Equation 6.16. Recall that the general *M*-estimator is the solution to

$$\sum_{i=1}^{n} \Psi(y; \theta) = 0, \qquad [6.17]$$

or, in the specific case of the generalized linear model,

$$\sum_{i=1}^{n} \Psi(y; \mu) = 0, \qquad [6.18]$$

where Ψ gives weight to the observations. As for *MM*-estimation for linear regression, if is odd and bounded, meaning $\rho(\cdot)$ is symmetric around 0, the breakdown point of the estimator is $BDP = 0.5$. Cantoni and Ronchetti accomplish this by solving

$$\Psi(y; \mu) = v(y; \mu)w(\mathbf{x})\mu' - a(\boldsymbol{\beta}), \qquad [6.19]$$

where

$$a(\boldsymbol{\beta}) = \frac{1}{2} \sum_{i=1}^{n} E[v(y_i; \mu_i)]w(\mathbf{x}_i)\mu_i' \qquad [6.20]$$

and the v_i and w_i are weight functions that consider the residuals and the hat values of the observations respectively. An adaptation of the Huber function ensures that the weights are robust to unusual y values

$$v_i(y_i; \mu_i) = \Psi(e_i)\frac{1}{V^{1/2}(\mu_i)}. \qquad [6.21]$$

Following the original Mallows GM-estimator for linear models, a possible choice for $w_i(\mathbf{x}_i)$ is $w_i(x_i) = \sqrt{1 - h_i}$. As we have already seen, however, this produces a low breakdown point, so the inverse of the robust distances is employed instead (recall the discussions in Chapter 4 on robust distances). The end result is an estimator that is efficient, has bounded influence, and is asymptotically normal. More important, it has been shown that inference from this model is much more reliable than from ordinary GLMs when the data are contaminated (see Cantoni and Ronchetti 2001).

EXAMPLE 6.1: Logistic Regression Predicting Vote for the Labour Party in Britain, 2001

This example uses data from the 1997–2001 British Election Panel Study (Heath, Jowell, and Curtice 2002). We concentrate only on respondents who participated in the final wave in 2001. After missing data are removed, the analytical sample size is 1,421. The goal is to assess the impact of the leader of the Labour Party, Tony Blair, on vote for his party during the 2001 British election. The dependent variable is vote for the Labour Party (coded 1) versus otherwise (coded 0). Evaluations of Blair were tapped with a five-point Likert item asking respondents how well they thought Blair was doing as prime minister (high values indicated a good job). The analysis also controls for age; gender; education (degree, some postsecondary, a-level, o-level, and none); social class (managers/professionals, routine nonmanual, self-employed, manual working class); retrospective sociotropic economic perceptions (a five-point scale, with high values indicating that the respondent felt the economy improved in the past year); and retrospective egocentric economic perceptions (a five-point scale, with high values indicating that the respondent felt his or her own personal economic position improved in the past year).[3] Both a regular logistic regression and a robust regression are fitted to the data.

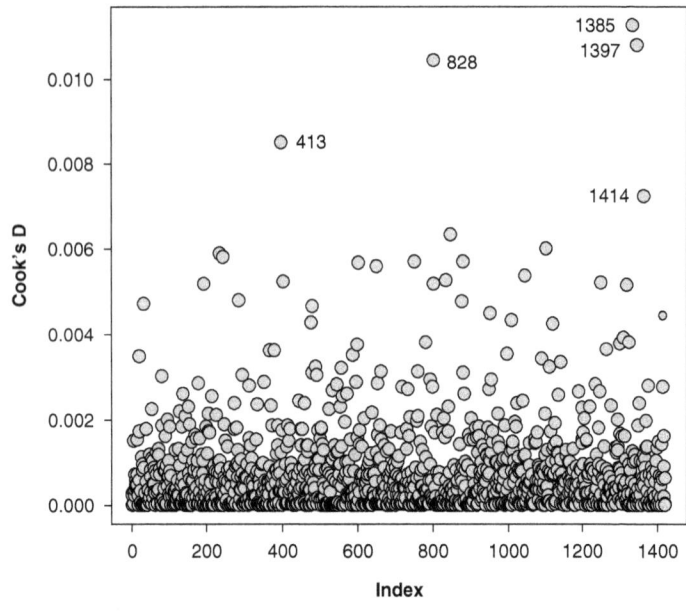

Figure 6.1   Index Plot of Cook's Ds for Logistic Regression Predicting Labour Vote in Britain, 2001

We start by assessing influence in the regular logit model. As we see from the index plot of Cook's Ds in Figure 6.1, a handful of observations have relatively high influence on the regression surface. Further diagnostics, including close examination of the $DFBETA_i$ for each coefficient in the model, failed to uncover any obvious problems, however. In other words, although some cases have unusually high influence overall, they do not appear to substantially influence any of the coefficients, at least not individually. Still, given the high overall influence of these cases, we explore whether or not a robust logistic regression tells a different story than the regular logistic regression.

Table 6.2 shows the results of the two regressions. Despite the presence of some observations with relatively high influence, the regular logistic regression has performed quite well. In fact, substantive conclusions are similar regardless for the two models—we would conclude that appraisals of Blair had a profound effect on whether or not someone voted for the Labour Party. Although the coefficient for the impact of appraisals of Tony Blair is slightly larger for the robust logistic regression (1.205 versus 1.127), the difference between the two coefficients is not statistically significant. The regular

TABLE 6.2
Logistic Regression Models Predicting Labour Vote in Britain, 2001

	Maximum Likelihood Logit Model		Robust Logit Model	
	$\hat{\beta}$	SE $\hat{\beta}$	$\hat{\beta}$	SE $\hat{\beta}$
Intercept	−5.15	0.468	−5.42	0.525
Age	−0.003	0.004	−0.003	0.004
Male	0.117	0.141	0.129	0.144
Education				
Degree	−0.372	0.269	−0.350	0.276
A-level	−0.391	0.255	−0.321	0.261
O-level	−0.190	0.181	0.163	0.187
Some post-sec	−0.462	0.235	−0.374	0.241
None	0	—	0	—
Social Class				
Professionals/managers	−0.055	0.184	−0.040	0.189
Routine nonmanual	−0.271	0.193	−0.213	0.197
Self-employed	−0.548	0.259	−0.551	0.266
Manual working class	0	—	0	—
Economic Perceptions				
Retrospective sociotropic	0.496	0.087	0.476	0.090
Retrospective egocentric	0.268	0.077	0.266	0.079
Opinions of Tony Blair	1.127	0.101	1.205	0.122
n	1,421		1,421	

logistic regression should be preferred for these data, then, because of its simplicity relative to the robust regression. This example is typical in that it is difficult for unusual observations to exert strong influence on the regression surface in logistic regression because the dependent variable can take on only two values. As we shall see below, however, unusual cases are more likely to exert high influence in Poisson regression.

EXAMPLE 6.2: Robust Poisson Regression Predicting Voluntary Association Membership in Quebec

This example uses data from the Canadian Equality, Security, and Community Survey of 2000. Although the data set contains information on respondents from across Canada, only Quebec respondents are included in the analysis ($n = 949$). The dependent variable is the number of voluntary associations to which respondents belonged. The independent variables are gender (with women as the reference category), Canadian born (the

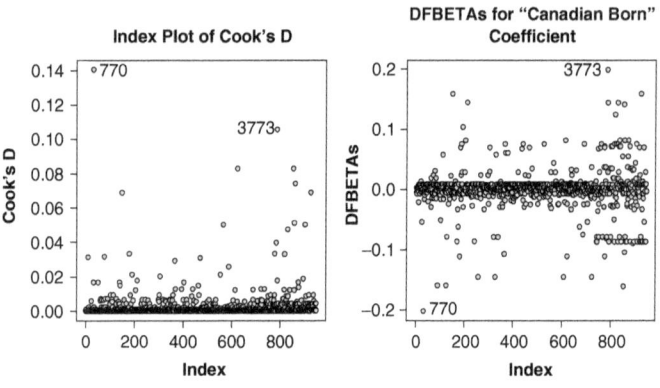

Figure 6.2   Diagnostic Plots for a Poisson Regression Model Predicting
Voluntary Association Involvement in Quebec

reference category is "not born in Canada"), and language spoken in the
home (divided into English, French, and other, with French coded as the
reference category). Given that the dependent variable is a count variable
(and follows a Poisson distribution), Poisson regression models are
employed. Both a regular generalized linear model using maximum likelihood and a robust GLM using quasi-likelihood are fitted. Before discussing
the results, we turn to diagnostic plots for the OLS regression.

Although extensive diagnostics were carried out, only those that uncov-
ered potentially problematic observations are reported. In this respect,
Figure 6.2 displays index plots for Cook's distances and the DFBETA$_i$ for
the "Canadian born" coefficient. The Cook's distances indicate that there
are perhaps 10 observations with fairly large influence on the regression,
two of which may be particularly problematic (observations 770 and 3773).
Analysis of the DFBETA$_i$ indicates that the influence of these two cases is
largely with respect to the effect of Canadian born, although as the plot indicates, their influences are in opposite directions.

Table 6.3 displays the results from the regular Poisson regression and the
robust Poisson regression. We see clearly that the coefficient for Canadian
born for the regular GLM was affected by unusual observations that did not
fit with the bulk of the data. The coefficient for the robust regression model
is nearly 10 times as large as the regular GLM coefficient. The difference in
effect makes for very different substantive interpretations. We would conclude from the regular GLM that, holding the other predictors constant,
there is no difference between those born in Canada and those born else-
where in terms of participation in voluntary associations ($e^{0.027} = 1.03$;

TABLE 6.3

Poisson Regression Models Predicting Voluntary
Association Membership

	Maximum Likelihood GLM			Robust GLM		
	$\hat{\beta}$	SE $\hat{\beta}$	$e^{\hat{\beta}}$	$\hat{\beta}$	SE $(\hat{\beta})$	$e^{hat\beta}$
Intercept	0.586	0.077	1.79	0.120	0.095	1.13
Men	0.079	0.045	1.08	0.084	0.053	1.09
Canadian born	0.027	0.072	1.03	0.258	0.088	1.29
Language						
English	0.357	0.061	1.43	0.537	0.068	1.71
Other	−0.014	0.094	0.98	0.079	0.112	1.08
French	0	0	1.00	0	0	1.00
n	949			949		

$p = .71$). On the other hand, the robust regression suggests that, on average, those born in Canada belong to 30 percent more associations at fixed values of the other predictors ($e^{0.258} = 1.29; p = .0035$).

The examples in this chapter are informative for two reasons. First, the Poisson regression example clearly showed that estimates from GLMs can be drastically altered by unusual observations. Conclusions based on the regular GLM were quite different from those based on the robust GLM. In this case, it makes the most sense to report the robust GLM. Second, the logistic regression example showed that even with a handful of observations with relatively high influence, the substantive conclusions from a robust GLM will not necessarily differ from those based on the regular GLM. Because the dependent variable can take on only two values—and hence it is usually impossible for the residuals to get extremely large—this is often the case for logistic regression. In these situations, the regular GLM is preferred because of its simplicity relative to the robust GLM. Nevertheless, it is worth exploring the robust GLM if only as a diagnostic tool.

Notes

1. An *M*-estimator is considered conditionally Fisher-consistent if

$$E_\beta[\Psi(y,x,\beta)|x] = \int\int \Psi(y,x,\beta)P_\beta(dy|x) = 0 \text{ for all } \beta \text{ and } x.$$

Maximum likelihood estimators for linear and generalized linear models are conditionally Fisher-consistent if the distribution of x does not depend on β.

2. This is the method employed by the glmrob function in the robustbase library for the statistical package **R**.

3. For more detailed information on the coding of the variables see Andersen and Evans (2003).

7. CONCLUSIONS

This book emphasizes the importance of detecting and properly handling unusual observations in regression analysis. The empirical examples demonstrate that if such cases go undetected, they could seriously distort the regression estimates. Evidence was also provided to indicate that vertical outliers, and more generally, heavy-tailed distributions can decrease the precision of regression estimates. These problems pertain to models fitted using both OLS and the more general GLM, and thus highlight the importance of diagnostics. Several traditional methods for detecting vertical outliers, leverage points, and influence were explored. Combined, these methods were effective in identifying problematic observations in the empirical examples.

Having identified problematic observations, the researcher can consider several options to accommodate them. The simple "fix" is to remove the offending observations from the analysis. This is a sensible strategy if there are good reasons for doing so, for example, if an observation is miscoded or known to be unique for some particular reason. Sometimes, however, the unusual observations reflect something systematic for which the model was unable to account. This is an important issue that implies that unusual observations are not always synonymous with "bad" data. In fact, the outliers could be the most intriguing part of the data. If there are many unusual observations, we should try to accommodate them by adding terms to the model—either new variables or interactions between existing variables—that account for the discrepancy. If no sound justification for the removal of the unusual observations or changes to the model specification can be determined, robust regression techniques are a suitable option.

On one hand, the strategy of robust regression is not much different from removing the observations. Both strategies have the goal of finding the best fitting model for the bulk of the data. In this respect, some might argue that both strategies result in a form of data truncation bias. In other words, by removing or down-weighting observations when we don't know if they are truly "contaminated" data, we are biasing the regression estimates. I disagree with this argument. We use statistical models to describe patterns in the data. The goal should be to tell the best possible "story" from the data. It doesn't make sense, then, to talk about a relationship between y and x, regardless of whether it is statistically significant or not, if it is driven completely by one or a few unusual observations. This would be a misleading story. Instead, we should be more concerned with patterns that fit the bulk of the data, not a few select observations. This does not mean, however, that we should automatically report robust regressions without considering why particular observations are unusual. In fact, I advocate using these methods primarily as a diagnostic tool, relying on them for final models only when we have no way of explaining the unusualness.

This book discussed many different types of robust regression, although most of the early methods should be considered largely obsolete. This extensive list was necessary only because more recent methods build on the earlier methods. Given that our main goal is to overcome the influence of unusual observations on the regression estimates, we should prefer estimators that have a high breakdown point and bounded influence. Many early robust regression methods, such as LMS, LTS, LTM, and S-estimation, satisfy these criteria. The problem with these methods is low efficiency under the Gauss-Markov assumptions. If we have the goal of making inferences from sample data to the larger population from which they were drawn, we should also desire efficient estimates. Robust regression methods that are based on M-estimation satisfy this last criterion. The original M-estimates, however, are not very resistant. Their breakdown point is 0, indicating that they can be seriously affected by even a single observation.

Fortunately, generalizations of M-estimates combine high resistance with high efficiency. These estimators accomplish this by starting with a very resistant fit (such as LMS, LTS, or S-estimation) to find initial estimates of the residuals and/or their scale, and then use this information in a final stage of M-estimation to minimize the residuals. The most important estimators to consider, at least if the robust regression will be used for the final model on which conclusions are made, are the GM-estimators of Coakley and Hettmansperger (1993), which are also known as one-step Schweppe's estimators, and the MM-estimators first proposed by Yohai (1987). Although the former perhaps best handle leverage points, they are very inefficient with small samples. The latter, on the other hand, perform reasonably well under most conditions. As was shown in Chapter 4, however, the early "resistant" but inefficient estimators can also play an important role when using robust regression as a diagnostic tool.

Despite the obvious attractiveness of estimators with a high breakdown point, they should be used with caution. A criticism of these techniques is that standard diagnostic methods have trouble detecting curvature (Cook, Hawkins, and Weisberg 1992; McKean, Sheather, and Hettmansperger 1993). Wilcox (2005) recommends that when high breakdown methods are employed, other methods that allow for better detection of curvature should be used to complement the robust regression. In this respect, nonparametric regression and generalized additive models (GAM) can be quite helpful. GAMs fall outside the scope of the present book, but I direct the reader to Hastie and Tibshirani (1990) for an extensive discussion of these models, and Fox (2000a, 2000b) for a good introductory treatment.

I hope this book makes it clear that we should not use any method blindly. Unless the data are very well behaved (i.e., the residuals are normal and there are no outliers), different regression techniques could give quite different answers depending on how the data are configured. A sensible strategy is to use both robust regression methods and OLS and their related diagnostics in the preliminary stages of data analysis. A simple comparison of the coefficients for these methods will often give a good indication of the effectiveness of the OLS regression. In the event that the estimates differ, RR-plots, which plot the residuals from the OLS regression against residuals from robust regressions, can give information about which observations cause the differences. Combining these diagnostics with traditional methods, such as Cook's distances and DFBETA, can be very helpful. In the end, the relative efficiency and simplicity of the OLS regression suggests that it should be preferred if it provides a reasonably good summary of the pattern in the data. The same principles apply with respect to the use of GLMs and their more robust alternatives.

APPENDIX: SOFTWARE CONSIDERATIONS FOR ROBUST REGRESSION

All of the statistical analyses presented in this book were carried out using **R** (R Development Core Team, 2006), an implementation of the **S** language. Aside from its superb functionality and flexibility, **R** is also attractive because it is available free of charge. **R** can be downloaded for various operating systems at http://cran.r-project.org. As well as the standard package and recommended packages that come with the basic **R** distribution, **R** has many add-on packages (called "libraries" in **S**) that are also free of charge. With a fast Internet connection, these are easily obtained and installed within **R** itself.

Although many of the methods employed in this book can be found in several packages, those listed below are very effective. The "MASS" package (associated with Venables and Ripley 2002) contains functions for *M, S,* and *MM*-estimation. The "robustbase" package (associated with Maronna et al. 2006) has functions for various robust regression methods—including *MM*-estimates, LTS, and robust generalized linear models—and some robust diagnostic plots based on the robust distances. Among other things, the "car" package (associated with Fox 2002) includes functions for traditional diagnostics for the linear and generalized linear model. The "boot" package (associated with Davidson and Hinkley 1997 but written by Angelo Canty) provides excellent generic bootstrapping functions that were used to construct the bootstrapped standard errors for the robust regressions that were reported. The "quantreg" package (associated with Koenker 2005) includes functions for fitting LAV regression. Finally, freely available **R** code associated with Wilcox (2005) includes many useful robust routines, including the *GM*-estimators discussed in the book. The data and **R** code used for all of the examples in this book can be downloaded from the book's Web site, www.sagepub.com/andersendata.

All of the packages discussed above are also available for SPlus. The most extensive package for robust regression in SPlus is the "robust" library, which allows for the fitting of *M*- and *MM*-estimates and other robust models for linear regression. It also allows fitting of various robust GLMs, including that associated with Kunsch et al. (1989). The "robust" package is also available for **R**, although unlike the other packages discussed above, it comes with a license fee from Insightful (the producers of SPlus).

SAS and Stata also have good facilities for robust regression, especially with respect to various *M*-estimators and *L*-estimators. The PROGRESS program in SAS has routines for LMS and LTS regression, and the

ROBUSTREG procedure (first available in Version 9) implements most of the commonly used robust regression techniques, including M-estimation, LTS, S-estimation, and MM-estimation. In Stata, the *rreg* command fits various robust regressions based on M-estimation, including MM-estimation, and the *qreg* command fits LAD and LAV regression. Bootstrapped standard errors are also easily calculated in Stata using the *bsreg* function. At present, robust GLMs are not available for either SAS or Stata.

Gauss also has options for robust regression, including LAD (and quantile regression more generally), and various types of M-estimation. Bootstrapped standard errors are automatically returned when a robust regression is fitted. Although less extensive in its robust regression facilities, LIMDEP can be used to fit LAD, LAV regressions, and some M-estimation, and to find bootstrapped standard errors for the estimates. Robust regression routines of any kind—whether for the linear model or the GLM—are nonexistent in SPSS.

REFERENCES

Andersen, R. and G. Evans. 2003. "Who Blairs Wins? Leadership and Voting in the 2001 Election." *British Elections & Parties Review,* 13: 229–47.

Andrews, D. F., P. J. Bickel, F. R. Hampel, P. J. Huber, W. H. Rogers, and J. W. Tukey. 1972. *Robust Estimates of Location.* Princeton, NJ: Princeton University Press.

Atkinson, A. C. 1985. *Plots, Transformations and Regression.* Oxford, England: Clarendon.

Atkinson, A. and M. Riani. 2000. *Robust Diagnostic Regression Analysis.* New York: Springer-Verlag.

Belsley, D. A., E. Kuhn, and R. E. Welsch. 1980. *Regression Diagnostics.* New York: Wiley.

Bianco, A. M. and V. J. Yohai. 1996. "Robust Estimation in the Logistic Regression Model." Pp. 17–34 in *Robust Statistics, Data Analysis, and Computer Intensive Methods,* edited by H. Rieder. New York: Springer-Verlag.

Bickel, P. J. and E. L. Lehmann. 1975. "Descriptive Statistics for Nonparametric Models II. Location." *Annals of Statistics* 3:1045–69.

Bickel, P. J. and E. L. Lehmann. 1976. "Descriptive Statistics for Nonparametric Models III. Dispersion." *Annals of Statistics* 4:1139–58.

Birch, J. B. and D. B. Agard. 1993. "Robust Inferences in Regression: A Comparative Study." *Communications in Statistics, Simulation and Computation* 22:217–44.

Birkes, D. and Y. Dodge. 1993. *Alternative Methods of Regression.* New York: Wiley.

Cantoni, E. and E. Ronchetti. 2001. "Robust Inference for Generalized Linear Models." *Journal of the American Statistical Association* 96:1022–30.

Carroll, R. J. and S. Pederson. 1993. "On Robustness in the Logistic Regression Model." *Journal of the Royal Statistical Society, Series B* 55:693–706.

Carroll, R. J. and A. H. Welsh. 1988. "A Note on Asymmetry and Robustness in Linear Regression." *American Statistician* 42:285–87.

Chatterjee, S. and A. S. Hadi. 1988. *Sensitivity Analysis in Linear Regression.* New York: Wiley.

Chave, A. D. and D. J. Thomson. 2003. "A Bounded Influence Regression Estimator Based on the Statistics of the Hat Matrix." *Journal of the Royal Statistical Society, Series C (Applied Statistics)* 52:307–22.

Chernick, M. R. 1999. *Bootstrap Methods: A Practitioner's Guide.* New York: Wiley.

Coakley, C. W. and T. P. Hettamansperger. 1993. "A Bounded Influence, High Breakdown, Efficient Regression Estimator." *Journal of the American Statistical Association* 88:872–80.

Cook, R. D. 1977. "Detection of Influential Observations in Linear Regression." *Technometrics* 19:15–18.

Cook, R. D. and D. M. Hawkins. 1990. "Unmasking Multivariate Outliers and Leverage Points: Comment." *Journal of the American Statistical Association* 85:640–44.

Cook, R. D. and S. Weisberg. 1982. *Residuals and Influence in Regression.* London: Chapman Hall.

Cook, R. D. and S. Weisberg. 1999. *Applied Regression Including Computing and Graphics.* New York: Wiley.

Cook, R. D., D. M. Hawkins, and S. Weisberg. 1992. "Comparison of Model Misspecification Diagnostics Using Residuals From Least Mean of Squares and Least Median of Squares Fit." *Journal of the American Statistician* 87:419–24.

Copas, J. B. 1988. "Binary Regression Models for Contaminated Data." *Journal of the Royal Statistical Society, Series B* 50:225–65.

Cramer, H. 1946. *Mathematical Methods of Statistics.* Princeton, NJ: Princeton University Press.

Croux, C., P. J. Rousseeuw, and O. Hossjer. 1994. "Generalized S-Estimators." *Journal of the American Statistical Association* 89:1271–81.

Davidson, A. C. and D. V. Hinkley. 1997. *Bootstrap Methods and Their Application.* Cambridge, England: Cambridge University Press.

Davis, J. B. and J. W. McKean. 1993. "Rank-Based Methods for Multivariate Linear Models." *Journal of the American Statistical Association* 88:245–51.

De Jongh, P. J., T. De Wet, and A. H. Welsh. 1988. "Mallows-Type Bounded-Influence-Regression Trimmed Means." *Journal of the American Statistical Association* 83:805–10.

Dietz, T. R., S. Frey, and L. Karloff. 1987. "Estimation With Cross-National Data: Robust and Nonparametric Methods." *American Sociological Review* 52:380–90.

Dobson, A. J. 1990. *An Introduction to Generalized Linear Models.* New York: Chapman Hall.

Draper, N. R. and H. Smith. 1998. *Applied Regression Analysis.* New York: Wiley.

Dunteman, G. H. and M. R. Ho. 2005. *An Introduction to Generalized Linear Models* (Quantitative Applications in the Social Sciences, 07–145). Thousand Oaks, CA: Sage.

Efron, B. 1979. "Bootstrap Methods: Another Look at the Jackknife." *Annals of Statistics* 7:1–26.

Efron, B. 1981. Nonparametric Standard Errors and Confidence Intervals (With Discussion)." *Canadian Journal of Statistics* 9:139–72.

Efron, B. and R. J. Tibshirani. 1993. *An Introduction to the Bootstrap.* New York: Chapman Hall.

Fahrmeir, L. and G. Tutz. 2001. *Multivariate Statistical Modelling Based on Generalized Linear Models.* 2d ed. New York: Springer.

Ferretti, N., D. Kelmansy, V. J. Yohai, and R. H. Zamar. 1999. "A Class of Locally and Globally Robust Regression Estimates." *Journal of the American Statistical Association* 94:174–88.

Fox, J. 1991. *Regression Diagnostics: An Introduction* (Quantitative Applications in the Social Sciences, 07–079). Newbury Park, CA: Sage.

Fox, J. 1997. *Applied Regression Analysis, Linear Models, and Related Methods.* Thousand Oaks, CA: Sage.

Fox, J. 2000a. *Simple Nonparametric Regression.* (Quantitative Applications in the Social Sciences, 07–129). Thousand Oaks, CA: Sage.

Fox, J. 2000b. *Multiple and Generalized Nonparametric Regression.* (Quantitative Applications in the Social Sciences, 07–130). Thousand Oaks, CA: Sage.

Fox, J. 2002. *An R and S-PLUS Companion to Applied Regression.* Thousand Oaks, CA: Sage.

Fung, W. K. 1999. "Outlier Diagnostics in Several Multivariate Samples." *Statistician* 48: 73–84.

Gill, J. 2001. *Generalized Linear Models. A Unified Approach* (Quantitative Applications in the Social Sciences, 07–134). Thousand Oaks, CA: Sage.

Hampel, F. R. 1974. "The Influence Curve and Its Role in Robust Estimation." *Journal of the American Statistical Association* 69:383–93.

Hampel, F. R. 1975. "Beyond Location Parameters: Robust Concepts and Methods." *International Statistical Institute, Proceedings of the 40th Session* 46:375–91.

Hampel, F. R., E. Z. Ronchetti, P. J. Rousseeuw, and W. A. Stahel. 1986. *Robust Statistics: The Approach Based on Influence Functions.* New York: Wiley.

Handschin, E., F. C. Schweppe, J. Kohlas, and A. Fiechter. 1975. "Bad Data Analysis for Power System State Estimation." *IEEE Transactions of Power Apparatus and Systems, PAS-94,* 329–37.

Hao, L. and D. Q. Naiman. 2007. *Quantile Regression* (Quantitative Applications in the Social Sciences, 07–149). Thousand Oaks, CA: Sage.

Hastie, T. J. and R. Tibshirani. 1990. *Generalized Additive Models.* London: Chapman Hall.

Hastie, T., R. Tibshirani, and J. Friedman. 2001. *The Elements of Statistical Learning: Data Mining, Inference and Prediction.* New York: Springer.

Heath, A., R. Jowell, and J. K. Curtice. 2002. *British Election Panel Study, 1997–2001: Waves 1 to 8 [computer file].* 4th ed. Colchester, Essex, England: UK Data Archive [distributor], July 2002. SN: 4028.

Hill, R. W. and P. W. Holland. 1977. "Two Robust Alternatives to Least Squares Regression." *Journal of the American Statistical Association* 72:828–33.

Hoaglin, D. C., F. Mosteller, and J. W. Tukey. 1983. *Understanding Robust and Exploratory Data Analysis.* New York: Wiley.

Hogg, R. V. 1974. "Adaptive Robust Procedures." *Journal of the American Statistical Association* 69:909–27.

Huber, P. J. 1964. "Robust Estimation of Location Parameters." *Annals of Mathematical Statistics* 35:73–101.

Huber, P. J. 1973. "Robust Regression: Asymptotics, Conjectures and Monte Carlo." *Annals of Statistics* 1:799–821.

Huber, P. J. 2004. *Robust Statistics.* Hoboken, NJ: Wiley.

Inglehart, R. et al. 2000. *World Values Surveys and European Values Surveys, 1981–84, 1990–93, and 1995–97* [Data file and codebooks]. Ann Arbor, MI: ICPSR.

Jaeckel, L. A. 1972. "Estimating Regression Coefficients by Minimizing the Dispersion of the Residuals." *Annals of Mathematical Statistics* 43:1449–58.

Jasso, G. 1985. "Marital Coital Frequency and the Passage of Time: Estimating the Separate Effects of Spouses' Ages and Marital Duration, Birth and Marriage Cohorts, and Period Influences." *American Sociological Review* 50:224–41.

Jasso, G. 1986. "Is It Outlier Deletion or Is It Sample Truncation? Notes on Science and Sexuality." *American Sociological Review* 51:738–42.

Jurečková, J. and J. Picek. 2006. *Robust Statistical Methods With R.* New York: Chapman Hall.

Jurečková, J. and P. K. Sen. 1996. *Robust Statistical Procedures: Asymptotics and Interrelations.* New York: Wiley.

Kahn, J. R. and J. R. Udry. 1986. "Marital Coital Frequency: Unnoticed Outliers and Unspecified Interactions Lead to Erroneous Conclusions." *American Sociological Review* 51:734–37.

Kempthorne, P. J. and M. B. Mendel. 1990. "Unmasking Multivariate Outliers and Leverage Points: Comment." *Journal of the American Statistical Association* 85:647–48.

Kenney, J. F. and E. S. Keeping. 1962. *Mathematics of Statistics.* 3d ed. Princeton, NJ: Van Nostrand.

Kim, S. J. 1992. "The Metrically Trimmed Mean as a Robust Estimator of Location." *Annals of Statistics* 26:1534–47.

Koenker, R. 2005. *Quantile Regression.* Cambridge, England: Cambridge University Press.

Koenker, R. W. and Bassett, G. W. 1978. "Regression Quantiles." *Econometrica* 46:33–50.

Koenker, R. W. and V. d'Orey. 1994. "Computing Regression Quantiles." *Applied Statistics* 43:410–14.

Krasker, W. S. and R. E. Welsch. 1982. "Efficient Bounded-Influence Regression Estimation." *Journal of the American Statistical Association* 77:595–604.

Kunsch, H. R., L. A. Stefanski, and R. J. Carroll. 1989. "Conditionally Unbiased Bounded-Influence Estimation in General Regression Models, With Applications to Generalized Linear Models." *Journal of the American Statistical Association* 84:460–66.

Lawrence, K. D. and J. L. Arthur, eds. 1990. *Robust Regression: Analysis and Applications.* New York: Marcel Dekker.

98

Lax, D. A. 1985. "Robust Estimators of Scale: Finite Sample Performance in Long-Tailed Symmetric Distributions." *Journal of the American Statistical Association* 80:736–41.

Leger, C. and J. P. Romano. 1990. "Bootstrap Adaptive Estimation: The Trimmed-Mean Example." *Canadian Journal of Statistics* 18:297–314.

Li, D. 1985. "Robust Regression." In *Exploring Data Tables, Trends and Shapes*, edited by D. Hoaglin, F. Mosteller, and J. Tukey. New York: Wiley.

Liao, T. F. 1994. *Interpreting Probability Models: Logit, Probit, and Other Generalized Linear Models* (Quantitative Applications in the Social Sciences, 07–101). Thousand Oaks, CA: Sage.

Lindsey, J. K. 1997. *Applying Generalized Linear Models*. New York: Springer-Verlag.

Long, J. S. and L. H. Ervin. 2000. "Using Heteroskedasticity Consistent Standard Errors in the Linear Regression Model." *American Statistician* 54:217–24.

Maronna, R. A., O. H. Butos, and V. J. Yohai. 1979. "Bias and Efficiency Robustness of General M-Estimators for Regression With Random Carriers." Pp. 91–116 in *Smoothing Techniques for Curve Estimation*, edited by T. Gasser and M. Rosenblatt. New York: Springer-Verlag.

Maronna, R. A., D. R. Martin, and V. J. Yohai. 2006. *Robust Statistics: Theory and Methods*. New York: Wiley.

McCullagh, P. and J. A. Nelder. 1989. *Generalized Linear Models*. 2nd ed. New York: Chapman Hall.

McKean, J. W. and S. J. Sheather. 2000. "Partial Residual Plots Based on Robust Fits." *Technometrics* 42:249–61.

McKean, J. W., S. J. Sheather, and T. P. Hettmansperger. 1993. "The Use and Interpretation of Residuals Based on Robust Regression." *Journal of the American Statistical Association* 88:1254–63.

McKean, J. W. and T. J. Vidmar. 1994. "A Comparison of Two Rank-Based Methods for the Analysis of Linear Models." *American Statistician* 48:220–29.

Mooney, C. Z. and R. D. Duval. 1993. *Bootstrapping. A Nonparametric Approach to Statistical Inference* (Quantitative Applications in the Social Sciences, 07–095). Newbury Park, CA: Sage.

Mosteller, F. and J. W. Tukey. 1977. *Data Analysis and Regression: A Second Course in Statistics*. Reading, MA: Addison-Wesley.

Myers, R. 1990. *Classical and Modern Regression With Applications*. 2d ed. Boston: Duxbury.

Naranjo, J. D. and T. P. Hettmansperger. 1994. "Bounded Influence Rank Regression." *Journal of the Royal Statistical Society, Series B (Methodological)* 56:209–20.

Nelder, J. A. and R. W. M. Wedderburn. 1972. "Generalized Linear Models." *Journal of the Royal Statistical Society, Series A* 135:370–84.

Newey, W. K. and K. D. West. 1987. "A Simple, Positive Semi-Definite, Heteroskedasticity and Autocorrelation Consistent Covariance Matrix." *Econometrica* 55:703–8.

Pena, D. and V. Yohai. 1999. "A Fast Procedure for Outlier Diagnostics in Large Regression Problems." *Journal of the American Statistical Association* 94:434–45.

Pregibon, D. 1981. "Logistic Regression Diagnostics." *Annals of Statistics* 89:705–24.

Preisser, J. S. and B. F. Qaqish. 1999. "Robust Regression for Clustered Data With Applications to Binary Regression." *Biometrics* 55:574–79.

R Development Core Team. 2006. *R: A Language and Environment for Statistical Computing*. Vienna, Austria: R Foundation for Statistical Computing. ISBN 3–900051–07–0, URL http://www.R-project.org.

Ramsay, J. O. 1977. "A Comparative Study of Several Robust Estimates of Slope, Intercept, and Scale in Linear Regression." *Journal of the American Statistical Association* 72:608–15.

Reed, J. F. 1998. "Contributions to Adaptive Estimation." *Journal of Applied Statistics* 25:651–69.

Rousseeuw, P. J. 1984. "Least Median of Squares Regression." *Journal of the American Statistical Association* 79:871–80.

Rousseeuw, P. J. 1985. "Multivariate Estimation With High Breakdown Point." Pp. 283–297 in *Mathematical Statistics and Applications, Vol. B*, edited by W. Grossman, G. Pflug, I. Vince, and W. Wetz. Dordrecht: Reidel.

Rousseeuw, P. J. and C. Croux. 1993. "Alternatives to the Median Absolute Deviation." *Journal of the American Statistical Association* 88:1273–83.

Rousseeuw, P. J. and A. M. Leroy. 1987. *Robust Regression and Outlier Detection*. New York: Wiley.

Rousseeuw, P. J. and W. C. van Zomeren. 1990. "Unmasking Multivariate Outliers and Leverage Points." *Journal of the American Statistical Association* 85:633–39.

Rousseeuw, P. J. and V. Yohai. 1984. "Robust Regression by Means of S-Estimators." *Nonlinear Time Series Analysis: Lecture Notes in Statistics* 26:256–72.

Ruppert, D. and D. G. Simpson. 1990. "Unmasking Multivariate Outliers and Leverage Points: Comment." *Journal of the American Statistical Association* 85:644–46.

Ryan, T. P. 1997. *Modern Regression Methods*. New York: Wiley.

Siegel, A. F. 1982. "Robust Regression Using Repeated Medians." *Biometrika* 69:242–44.

Stefanski, L. A., R. J. Carroll, and D. Ruppert. 1986. "Optimally Bounded Score Functions for Generalized Linear Models With Applications to Logistic Regression." *Biometrika* 73: 413–24.

Stromberg, A. J. 1993. "Computation of High Breakdown Nonlinear Regression Parameters." *Journal of the American Statistical Association* 88:237–44.

Stromberg, A. J., O. Hossjer, and D. M. Hawkins. 2000. "The Least Trimmed Differences Regression Estimator and Alternatives." *Journal of the American Statistical Association* 95:853–64.

Tukey, J. W. 1991. "Graphical Displays for Alternative Regression Fits." Pp. 309–26 in *Directions in Robust Statistics and Diagnostics, Part 2*, edited by W. Stahel and S. Weisberg. New York: Springer-Verlag.

Venables, W. N. and B. D. Ripley. 2002. *Modern Applied Statistics With S*. 4th ed. New York: Springer.

Weakliem, D. L., R. Andersen, and A. F. Heath. 2005. "By Popular Demand: The Effect of Public Opinion on Income Inequality." *Comparative Sociology* 4:261–84.

Wedderburn, R. W. M. 1974. "Quasilikelihood Functions, Generalized Linear Models and the Gauss-Newton Method." *Biometrika* 61:439–47.

Western, B. 1995. "Concepts and Suggestions for Robust Regression Analysis." *American Journal of Political Science* 39:786–817.

White, H. 1980. "A Heteroskedasticity-Consistent Covariance Matrix Estimator and a Direct Test for Heteroskedasticity." *Econometrica* 48:817–38.

Wilcox, R. R. 2005. *Introduction to Robust Estimation and Hypothesis Testing*. New York: Elsevier Academic Press.

Wu, L. L. 1985. "Robust M-Estimation of Location and Regression." In *Sociological Methodology 1985*, edited by N. Tuma. San Francisco: Jossey-Bass.

Yohai, V. J. 1987. "High Breakdown Point and High Efficiency Robust Estimates for Regression." *Annals of Statistics* 15:642–56.

Yohai, V. J. and R. H. Zamar. 1988. "High Breakdown-Point Estimates of Regression by Means of the Minimization of an Efficient Scale." *Journal of the American Statistical Association* 83:406–13.

INDEX

Voluntary association membership, 87–89

Weakliem, D. L. R., 24, 26
Weisberg, S., 37, 43
Welsch, R. E., 41
Wilcox, R. R., 93

Wilcoxon scores, 50
Working residuals, 82
World Values Survey, 24

Yohai, V. J., 54, 55, 56, 70

Zamar, R. H., 56

ABOUT THE AUTHOR

Robert Andersen is a professor of sociology and political science at the University of Toronto, Canada. His research interests are in applied statistics, political sociology (especially the social bases of attitudes and political behavior), social stratification, and the sociology of work. Some of his recent work has appeared in the *American Sociological Review, The Journal of Politics*, and *Sociological Methodology*.

www.ingramcontent.com/pod-product-compliance
Lightning Source LLC
Jackson TN
JSHW080037010226
97517JS00015B/158